Epistemic Ecology

Catherine Z. Elgin

The MIT Press
Cambridge, Massachusetts
London, England

The MIT Press
Massachusetts Institute of Technology
77 Massachusetts Avenue, Cambridge, MA 02139
mitpress.mit.edu

© 2025 Massachusetts Institute of Technology

This work is subject to a Creative Commons CC-BY-NC-ND license.

This license applies only to the work in full and not to any components included with permission. Subject to such license, all rights are reserved. No part of this book may be used to train artificial intelligence systems without permission in writing from the MIT Press.

The MIT Press would like to thank the anonymous peer reviewers who provided comments on drafts of this book. The generous work of academic experts is essential for establishing the authority and quality of our publications. We acknowledge with gratitude the contributions of these otherwise uncredited readers.

This book was set in Stone Serif and Stone Sans by Westchester Publishing Services. Printed and bound in the United States of America.

Library of Congress Cataloging-in-Publication Data

Names: Elgin, Catherine Z., 1948- author.
Title: Epistemic ecology / Catherine Z. Elgin.
Description: Cambridge, Massachusetts : The MIT Press, [2025] | Includes bibliographical references.
Identifiers: LCCN 2024017233 (print) | LCCN 2024017234 (ebook) | ISBN 9780262551717 (paperback) | ISBN 9780262382403 (epub) | ISBN 9780262382410 (pdf)
Subjects: LCSH: Knowledge, Theory of. | Ecology—Philosophy. | Praxeology.
Classification: LCC BD161 .E444 2025 (print) | LCC BD161 (ebook) | DDC 121—dc23/eng/20240703
LC record available at https://lccn.loc.gov/2024017233
LC ebook record available at https://lccn.loc.gov/2024017234

10 9 8 7 6 5 4 3 2 1

EU product safety and compliance information contact is: mitp-eu-gpsr@mit.edu

For Jim and Sam and Gareth with love

Contents

Acknowledgments   ix

1   Epistemic Agency   1
2   Epistemic Autonomy   27
3   Epistemic Engagement   59
4   Epistemic Dynamics   81
5   Realms of Epistemic Ends   121
6   Word Giving   141
7   Word Taking   163
8   Reasonable Disagreement   177
9   Beyond the Information Given   211
10   Constructive Nominalism   241
11   Models as Felicitous Falsehoods   263
12   Epistemic Gatekeepers   283

Notes   305
References   313
Index   325

Acknowledgments

Many friends and colleagues, as well as a few virtual strangers, helped me with this book. Christoph Jäger and Johanna Stüger organized a workshop at Humboldt Universität where the manuscript was discussed and dissected. Glen Andrenau, Joseph Bjelde, Fabio Lampert, Fredrica Malfatti, Christopher Ranalli, Nicholas Shakel, and Michael Vollmer took part. Their advice was invaluable. Johanna then sent me incisive comments on the entire manuscript along with an embarrassingly long list of typos. Claus Beisbart, Georg Brun, and Tanja Rechnitzer hosted a workshop at the University of Bern on reflective equilibrium. The discussions of the participants were invaluable. Jonathan Matheson and Kirk Lougheed sent comments that improved the section on epistemic autonomy. Kirk also deepened my thinking about disagreement, as did Casey Johnson. Dorothy Edgington raised useful worries about the Kantian bent of the project. Insa Lawler and Finnur Dellsén provided insights and telling objections to my views about epistemic progress. Randy Curran and Harvey Siegel were invaluable critics of my views on the philosophy of education. Harvey made helpful comments on the rest of the manuscript as well. Jan-Willem Romeijn raised issues about objectivity. Otávio Bueno, Anjan Chakravartti, Henk de Regt, Steven French, Roman Frigg, Kareem Khalifa, Doug Marshall, and Nancy Nersessian helped me with issues in the sections on philosophy of science. Dom Lopes and Chiara Ambrosio deepened my understanding of the powers and limits of various modes of representation. From different perspectives, Devika Agrawal, Sam Elgin, Mike Hannon, Federica Malfatti, and Paul Teller, raised issues about pretty much everything. I am grateful to you all. Some of my closest friends and most acute critics are no longer with us: Jonathan Adler, Judy Thomson, Amélie Rorty. Their influence pervades the book.

Ancestors of various sections have been previously published. A version of part of chapter 2 first appeared as "The Realm of Epistemic Ends" in *Epistemic Autonomy*, edited by Jonathan Matheson and Kirk Lougheed (Oxford: Oxford University Press [2022], pp. 55–70). An early version of chapter 6 appeared as "Word Giving, Word Taking" in *Fact and Value: Essays for Judith Jarvis Thomson*, edited by Alex Byrne, Robert Stalnaker, and Ralph Wedgwood (Cambridge, MA: MIT Press [2001], pp. 97–116). I am grateful to the MIT Press for permission to reprint it. Chapter 8 is a medley, involving elements of "Disagreement in Philosophy," which originally appeared in *Synthese* (vol. 200, 20 [2022]); "Reasonable Disagreement" in *Voicing Dissent*, edited by Casey Johnson (Routledge [2018], pp. 10–21); and "Begging To Differ" in *The Philosopher's Magazine* (vol. 59 [2012], pp. 77–82). A section of chapter 9 was previously published as "Understanding as an Educational Objective" in *The Handbook of the Philosophy of Education*, edited by Randall Curren (London: Routledge [2022], pp. 69–78). Another section was published as "Beyond the Information Given: Teaching, Testimony, and the Advancement of Understanding" in *Philosophical Topics* (vol. 49 [2021], pp. 17–34; Copyright © 2021 by the Board of Trustees of the University of Arkansas. Reprinted with the permission of The Permissions Company, LLC on behalf of the University of Arkansas Press. All rights reserved). An earlier version of chapter 10 was published as "Nominalism, Realism, and Objectivity" in *Synthese* (vol. 196 [2019], pp. 519–534); an earlier version of chapter 11 as "Models as Felicitous Falsehoods" in *Principia* (vol. 26 [2022], pp. 7–23); and an earlier version of chapter 12 as "Epistemic Gatekeepers: The Role of Aesthetic Factors in Science" in *The Aesthetics of Science: Beauty, Imagination, and Understanding*, edited by Steven French and Milena Ivanova (London: Routledge [2020], pp. 21–35). I am grateful to all the editors, publishers, and anonymous referees for their contributions to these works.

My debts to my family, friends, colleagues, and students are incalculable. I thank you all.

1 Epistemic Agency

Waiting for Godot was not a viable strategy. If Vladimir and Estragon wanted to find Godot, they should have gone looking. They should have figured out what approaches to take, what methods to use, what instruments to deploy, what evidence would indicate that they were (or were not) on the right track. They should have learned from their mistakes, recognizing which approaches were fruitless or counterproductive and which ones held promise. If available methods or instruments proved inadequate, they should have devised better ones. They should have recruited allies and used their diverse talents and areas of expertise. That is what effective epistemic agents do. There is no guarantee of success, but taking the initiative is better than sitting around hoping that epistemic success will fall into one's lap.

The Epistemic Predicament

Human beings are cognitively limited, error prone, and inveterately curious. There is plenty we do not know, plenty we do not understand, plenty we are mistaken about. Some of our epistemic inadequacies are glaringly obvious; others are hidden from view. Often we are capable of finding out or figuring out the answers to our questions. We pull out our smartphones, consult an expert, make a calculation, or conduct a test whose results we rightly consider reliable. But sometimes our epistemic resources prove inadequate. Our best efforts fall short. They do not afford access to the phenomena we seek to understand, or the experts are at odds, or the available techniques do not equip us to identify or correct flaws in our convictions or methods. Even worse, we may be at a loss about how even to approach the quandaries we confront. This is the human epistemic predicament. Our reach exceeds our grasp. We relentlessly seek to expand our epistemic range—to find out things we do not already know, grasp matters we do not already understand, correct errors and other gaffes in our thinking, compensate for our cognitive limitations. Some of our quests are momentous; others, exercises in idle curiosity. But being inquisitive, we humans relentlessly attempt to improve our epistemic lot.

Put this way, the epistemic predicament might seem to plague only an intellectual elite—astrophysicists investigating dark matter, philosophers working on the problem of free will, historians attempting to discern the political machinations at the Congress of Vienna. In fact, however, everyone regularly confronts and seeks to alleviate their own cognitive limitations. One person wants to understand why the Celtics so frequently squander their lead; another wants to identify the bug that is eating the begonias; a third wants to account for the popularity of torn jeans. When epistemic impediments reflect purely personal limitations, we can readily resolve them by appealing to a trustworthy source. By consulting a field guide to insects or asking an entomologist, the curious gardener could probably identify the begonia-eating bug. Other limitations are general. Perhaps the reasons for the team's inability to sustain a lead are complicated and not fully understood by anyone. Perhaps fashionistas diverge over the reasons for fashion trends, and there is no clear basis for favoring one expert opinion over another.

When epistemic inadequacies are sufficiently intriguing, irksome, important, or severe, we develop techniques and approaches to alleviate them. In attempting to improve our epistemic lot, it won't do simply to identify the end state—to say, for example, that since knowledge consists in non-fortuitously justified true belief or whatever, an agent knows when and only when she satisfies this requirement. We need to determine how to get there, given that we start from considerations that are not themselves entirely well founded. We also need to determine how to recognize that we have arrived. How do we tell that we've satisfied the requisite requirement? These are not idle questions, for we want—indeed, need—to exploit our epistemic achievements. The value of such achievements lies in their equipping us justifiably to think and act confidently on the basis of the support they provide. Merely satisfying the conditions on knowledge or understanding without having reason to think you satisfy them is insufficient. Even if the bridge is safe, you would be unwise to cross it unless you had reason to think it safe. Even if the test is reliable, you would be ill-advised to count on it unless you had reason

Epistemic Agency

to consider it reliable. It is not enough to base our convictions on considerations that are in fact sound. We need to be equipped to assess their soundness. This sets an additional epistemic demand. We must be able to evaluate the scope and limits of our convictions and our justification for them. We need to understand ourselves as epistemic agents. Moreover, we need to be able to tell whether and to what extent our epistemic situation is precarious. When it is, we should shore up our defenses, take precautions, hedge our bets. And we need to know how to do this as well.

Epistemic Ecology inquires into the ways finite and fallible people individually and collectively develop increasingly viable understandings. Its thesis is that autonomous epistemic agents extend, modify, and adapt to their environments, both social and natural, in order to frame and promote their epistemic ends. To make this out requires an explication of autonomous epistemic agency, an account of the constitution of epistemic communities, and an explanation of the ways such agents and communities both modulate and adapt to the sometimes recalcitrant phenomena they seek to understand. I argue that our epistemic orientation is agential, not spectatorial. Epistemic subjects are agents, not mere observers; the positions they accept are springboards for improvement rather than mere windows on the world. This raises the issue of epistemic dynamics. Rather than treating epistemic success—knowledge, understanding, wisdom, or whatever—as something fixed and final, I construe success as a suitably stable platform to build on. I ask how we should leverage our findings to move beyond them. In this chapter, I present a schematic overview of the position, highlighting themes that will be addressed in greater detail in subsequent chapters.

Fallibility and finitude might seem to be regrettable handicaps. We contrive methods to overcome the limits they impose and to protect ourselves against the vulnerabilities they engender. But, we may think, it would be preferable if we did not have to. I will argue otherwise. Properly deployed, epistemic obstacles transmute into epistemic opportunities. If bones never broke, orthopedic science would be much more limited than it is—not only because scientists would have less incentive to study bones, but also because the understanding of when and how and with what consequences bones break embeds information about the nature, structure, and function of bones and other anatomical tissue. Medical scientists are in a position to understand human anatomy and physiology better as a result. Similarly, if we never made mistakes (or a certain sort of mistakes), we would lack information about human reasoning that our errors and our making of those errors reveal. The errors may stem from limitations in our psychological makeup, our conceptions of the phenomena, or our standard modes of reasoning about them. The widespread propensity to make specific sorts of reasoning errors affords insight into human psychology (see Tversky & Kahneman 1974). We understand the mind better as a result of investigating the propensity to make such errors, and we understand it even better when we attend to our increasingly successful efforts to avoid and correct such errors.

Hand-waving about fallibility and finitude in general accomplishes little. But when we appreciate what specifically is wanting and why, we are in a position to leverage our epistemic achievements, to compensate for our limitations, to devise more powerful epistemic resources, and to deepen our understandings of ourselves, our compatriots, and our communities as epistemic agents. We improve and expand epistemic access by inventing instruments and techniques to disclose what unaided perception cannot. Galileo could see the moons of Jupiter because he refined the design of the telescope, enhancing its magnification. His innovation was not merely a technical achievement. Besides improving the design of the instrument, he recognized how the telescope could be used to extend his perceptual range and how the deliverances of his enhanced perception enabled him to augment and correct received views about the cosmos (see Galilei 2015). This required developing not just an artificial lens but also an understanding of the powers and possibilities that artificial lenses afford.

This book is an investigation into epistemic ecology: the relation of individual epistemic agents to their natural and social environments that enables them, with some measure of success, to form and pursue their epistemic objectives. Like any ecology, it is a story of mutual adjustments between organisms and environments, where each adapts to the other. Humans are social animals. Part of a social animal's adapting to its environment consists in adapting to the behavior of its conspecifics. Similarly, part of human epistemic ecology consists in the individual's response to other members of her epistemic community.

Very schematically, the picture looks like this. An individual inquirer recognizes that she is finite and fallible. She has questions that, on her own, she cannot answer. She makes mistakes that, on her own, she cannot correct and sometimes cannot even recognize. She thus seeks to expand her range and

Epistemic Agency

remedy her inadequacies. Nor is she peculiar in this regard. Other inquirers face similar difficulties. But they are not clones of one another. Some have epistemic assets that others lack. They therefore join forces with the aim of overcoming individual cognitive limitations and exploiting collective cognitive strengths. They develop publicly accessible and assessable resources to achieve their ends—to check results, regiment reasons, stabilize findings. To do this, they need, and recognize that they need, common standards, methods, modes of representation, and metrics. They therefore contrive them and develop higher-order standards to assess their adequacy.

They do not always succeed, for the world pushes back. Despite the consensus on its desirability, they cannot build a perpetual motion machine. Nor can they square the circle. They revise their methods, standards, and goals in light of their setbacks. Even though they cannot give an exact numerical measure of the area of a circle, they can devise mathematical methods that generate ever better approximations. Even though they cannot construct a perpetual motion machine, they can invent machines that are increasingly efficient.

This suggests that the epistemic predicament is how finite and fallible agents can access a fixed and potentially recalcitrant external world. It is, of course, true that the phenomena we seek to understand often resist our efforts. We cannot individually or collectively ensure that things are as we take them to be. But that is just part of the story.

Bernard Williams describes that which is external to the mind as *what is there anyway* (1985:138). It presumably consists of the animals, vegetables, and minerals that constitute the natural world. *What is there anyway* is taken to be the object of empirical inquiry. But this Cartesian partition of phenomena into things that are constructions of the mind and things that are there anyway is misleading. Much that is external to the individual mind is nonetheless a product of human ingenuity: artifacts, which humans have created; institutions, whose powers and purposes humans have delineated; practices, which humans have constituted; games, whose rules and strategies humans have devised. They would not have existed without human activities. They were not there anyway.

Nor are we entirely passive onlookers with respect to natural phenomena. Epistemic agents regularly and effectively reorganize domains to answer to their interests. By inventing the calendar, ancient Sumerians reconfigured the endless sequence of days into repeatable units. Starting from phenomenal reports as of warmer and cooler, scientists devised not just a measuring instrument—the thermometer—but also the magnitude to be measured—temperature (Chang 2004). Such human innovations create mind-independent, often irritatingly recalcitrant, objective facts. The innovations do not always behave as their inventors anticipated. Once they exist, they are independent of our will. We devise methods to enable us to understand such things and to understand how they answer to, transcend, or diverge from their initially intended design.

Thinkers even create new epistemic realms, constituting new phenomena to be understood. Metrology, the science of measurement, concerns the accuracy, adequacy, and range of measuring instruments that humans have devised, the metrics for measurement that we have delineated, and the conditions that need to be satisfied for a phenomenon to be measurable. The phenomena metrology studies are products of human ingenuity, designed to foster epistemic ends. There are no measurements in the wild (Tal 2020). Much metallurgy concerns the nature, structure, and behavior of alloys—metals that humans invented. Bronze, brass, and steel have fixed, determinate, sometimes surprising, often refractory properties. Although alloys are human artifacts, once they are made, they and their properties are independent of the will. History, philosophy, and the social sciences study human institutions. Human beings devise methods, devices, standards of acceptability that frame and circumscribe their practices. Investigators seek to understand what they do and with what effects. Game theory studies reasoning strategies of rational agents. It equips us to understand both how we function as rational agents and where we fall short. Computer science studies the design and operation of algorithms, some of which were created by the computers themselves. The outputs studied far outstrip what was understood about their inputs. Still, both the computers and the initial algorithms are human artifacts. And the computer-generated algorithms are artifacts of human artifacts. The phenomena such fields investigate are products of epistemic efforts. They focus on how humans enhance and deepen their capacity to understand. None of this is surprising, but it does undermine the idea that understanding "the external world" is a matter of properly reflecting what is there anyway. To a considerable extent, the external world is as it is because of what we and our forebears have done.

Mutual Adaptations

Construing relations of an individual agent, an epistemic community, and the phenomena under investigation in ecological terms highlights mutual adaptations. The understandings that emerge are products of elaborate feedback loops. Using available epistemic resources, we revise what we think and how we investigate and assess, and we update and refine our resources on the basis of what we have found. The study of optics enables scientists to build lenses that broaden and improve their capacity to study optics. We recalibrate our thinking as well as our instruments. We restructure our epistemic communities as our understanding progresses. This results in increased specialization and a division of cognitive labor. The benefits are considerable, but the dependence on specialists increases individual vulnerability. Each of us has to rely on considerations that we cannot check for ourselves. So the moral and epistemic ties that bind the community of inquiry need to be reinforced as specialization grows. We are in a position to understand things better only if we have reason to trust our intellectual compatriots. Since many of the experts we rely on are strangers to us, we must devise and implement institutional bases for trust.

To constitute herself as an epistemic agent, a person has to constrain and regiment her thinking, holding herself responsible to standards that not all of her ideas meet. She does not credit every thought that runs through her head. She discounts dreams, idle fantasies, premonitions, wishful thinking. She takes hunches with a large grain of salt. She downgrades the strength of her convictions when she suspects that her evidence is inadequate or her orientation is skewed. She subjects her judgments to such restraints because she has learned that not every passing thought is worthy of credence. Drawing on the feedback of others, she fine-tunes her assessment of which of her ideas are creditable. She internalizes the norms of an epistemic community and subjects herself to them. She does so because she has reason to think that the considerations that satisfy those norms are far more likely to be creditable than those that do not. This is an effective procedure only to the extent that satisfying the community standards actually leads to more creditable results than flouting them or ignoring them would. That depends on what those standards are. For the community to perform its function, it must consist of members whose collective verdicts promote and are seen to promote their epistemic ends. An epistemic community is not a disorderly rabble; it emerges and structures itself precisely to serve epistemic ends.

Methods, metrics, and standards devised and validated by an epistemic community determine what the community counts as affording genuine access to its domain. The epistemic process that emerges is iterative. Investigations produce results that provide a platform for the construction of techniques and devices that enable further, more refined investigations that produce more and better results. The process incorporates mechanisms for self-scrutiny and self-correction, since innovations are not guaranteed to succeed. When they fail, it is often possible to discern when, where, and why they did so. This affords increased access to ourselves as epistemic agents, to our community as a monitor and enabler, and to the phenomena themselves. We get things wrong. But even though our best efforts go awry, we have the capacity to learn from our mistakes. So a failure need not be a total failure. And if a mistake was not a bone-headed blunder, the failed attempt may provide a back door to future success. We design our investigations to maximize our prospects, not only of success but, in default of that, to learn from our failures. This is not a Pollyannaish attempt to put a positive spin on our epistemic inadequacies. It is to highlight the fact that we intentionally contrive our institutions, practices, and devices in such a way that we can learn from our mistakes.

Reflecting Reality?

Richard Rorty (1979) maintains that the picture of the mind as the mirror of nature is a shibboleth of analytic philosophy. It is not a way that ordinary people, (that is, people who are not analytic philosophers) think about belief and knowledge. His argument is unpersuasive. The idea that knowledge requires truth and that truth consists in accurately reflecting the facts is both widespread and plausible. Stephen Grimm (2012:109–111) maintains that understanding too involves mirroring. Understanding, he contends, is more valuable than mere knowledge because it mirrors reality more deeply than mere knowledge does. To be sure, most people neither have nor think they need a fully developed, articulate epistemology. But it is common to rescind a claim to knowledge on learning that the facts are otherwise. "I thought I knew where I left my keys, but I was wrong." "I thought I knew who won the World Cup, but I didn't; it was Argentina, not France." We

Epistemic Agency

reproach informants who purported to know but turned out to be wrong. "You shouldn't have said it, since you didn't know." Such criticism seems apt only if a condition on knowledge is properly answering to the facts (see Stroud 1984:61–63). In chapter 10, I argue that this is a matter of refraction rather than reflection. Still, there is, as far as I can see, nothing odd or outré about taking knowledge to require reflecting the facts.

Much of what is epistemically creditable does not qualify as knowledge (see Elgin 2017). Scientific models, for example, are intentionally inaccurate representations of their targets. Since they are not true, they do not satisfy the requirements on knowledge. They are, I have argued, acceptable when they are true enough for the epistemic purposes for which they are used in the contexts in which they are used. Even so, it is not implausible to construe models that have targets as reflections.[1] Michael Strevens (2008:69–82) argues that a model is epistemically acceptable only if the elements that are difference makers reflect the facts. Quasi-factivists, such as Kareem Khalifa (2017:166–183) and Stephen Grimm (2014:336–338), might be read as maintaining that models provide slightly distorting mirrors of nature. This perspective can be extended beyond understanding of individual facts to comprehend understanding of subject matters as well. Then understanding consists in representing broad swaths of phenomena without (much) distortion. I do not want to enter into the details of these debates here. I mention them because the idea that epistemic success consists in mirroring the world seems to capture something plausible about what we take our epistemic situation to be. It is far from a mere watchword of analytic philosophy. Somehow, epistemic success involves answering to the facts.

Nevertheless, there is an element of the familiar characterization that has not been emphasized. The conceptions of knowledge and understanding that emerge are *spectatorial*. This is no idiosyncrasy of the positions I mention. Epistemology traditionally frames its subject matter in terms of the idea that the thinker stands to the phenomena as a spectator stands to what she sees. She observes it. And she is epistemically successful only if her representation of the facts accurately (enough) reflects those facts.

On such a conception, epistemic success is unconcerned with what particular facts are known or what constellation of facts is understood. A mirror, after all, indiscriminately reflects whatever is put in front of it. A distorting mirror distorts indiscriminately. To be sure, more than accuracy is required for epistemic success. A theory of knowledge sets standards for reliability or justification or epistemic virtue or standing in the proper relation to the facts. But the spectatorial stance is unconcerned about relevance, significance, and susceptibility to elaboration or correction. With suitable backing, the most trivial bit of information or network of information passes epistemic muster. A spectatorial conception of epistemic success answers "What do we know?" in a way that is indifferent to "Why should we care?" It is equally unconcerned with fruitfulness. A fruitful conclusion is one that provides a basis for further epistemic success; a sterile one does not. Assuming that epistemic closure holds, we know the obvious consequences of anything we know. We can thus validly infer infinitely many trivial truths from any truth we happen to know. From p, we can infer p *or* q and p *or* q *or* r and so forth ad infinitum. Because a disjunction is true if one of its disjuncts is, once we've ascertained that p is true, we need not concern ourselves with the truth values of further disjuncts. Infinitely many truths immediately follow. But because such inferences are trivial and their results are uninformative, they do virtually nothing to improve our epistemic lot. The fruitfulness of a finding is a function of its contribution to nontrivial inferences.

The spectatorial stance in epistemology yields a static conception of epistemic success. If at time t an individual satisfies the conditions on knowing that p, then at time t, she knows that p. If at time t, she satisfies the conditions on understanding φ, then at time t, she understands φ. Where, if anywhere, her current epistemic success leads is irrelevant. This might seem reasonable. Perhaps questions about the utility of a bit of knowledge or understanding, being practical, transcend the boundaries of epistemology per se. The knowledge that spinach is a good source of iron may be useful for improving nutrition or just another idle factoid that could someday figure in the solution to a crossword puzzle. The knowledge that it takes two and a half hours to drive from Cambridge to New Haven is the same whether it is used as a reason to go to New Haven or as a bit of trivia about distances between cities with Ivy League schools. Whether a person knows a particular fact is one thing; whether the fact is worth knowing is another.

I agree that considerations that satisfy spectatorial epistemic demands are genuine epistemic accomplishments. Spectatorial knowledge is genuine knowledge. Nor do I deny that in some cases—such as deciding to drive or to refrain from driving from Cambridge to New Haven or deciding to serve or refrain from serving spinach—epistemology has done its job when it vindicates a suitably supported reflection of the facts. But by adopting a

Epistemic Agency

spectatorial stance, epistemology unduly limits its purview. It does not position itself to promote the advancement of understanding or the growth of knowledge. Much of our knowledge stems from nontrivial inferences from what we already knew or reasonably believed. Much of what we understand is a product of judicious gleaning—identifying and retaining the epistemically valuable elements of a less than acceptable account while setting aside the chaff. Much is a product of revising commitments that are not quite right, of drawing and evaluating nontrivial inferences and so forth. Perhaps epistemology's job is done when it equips agents with the information needed to promote non-epistemic ends—to drive to New Haven or serve spinach. But, I suggest, its job is *not* done if the question concerns the promotion of *epistemic* ends, for that question cannot be dismissed as merely practical or prudential. It concerns how to exploit our epistemic achievements. The compensation for being epistemically limited and fallible is that we are capable of pushing beyond current epistemic limits and capable of correcting current errors.

The agential epistemic stance is dynamic; it is oriented toward its own improvement. Rather than focusing exclusively to what our current epistemic resources equip us to apprehend here and now, agential epistemology incorporates a concern with advancing understanding. L. Jonathan Cohen (1992:4) distinguishes between belief and acceptance. To believe that p is to be disposed to feel that p is so; to accept that p is to be willing (and, I add, able) to use p as a basis for inference or action when one's ends are cognitive. In earlier work, I argued that acceptance rather than belief is epistemically central (Elgin 2017). My original motive was to accommodate epistemically valuable models and idealizations that are not true hence ought not be believed. But there is more to the difference between acceptance and belief than that. Acceptance consists in an ability and willingness to *make* inferences and *perform* actions— that is, to *do* something. It therefore involves an orientation to the future. It construes epistemic subjects as agents, not onlookers. It equips us to recognize fruitfulness as a major epistemic value. Instead of taking epistemology to have done its job when it rules on whether particular judgments of fact are true and worthy of belief, it takes our epistemic objective to involve arriving at well-founded conclusions that we can build on. An acceptable commitment should be a constitutive component of a platform for further epistemic advances. Belief, in Cohen's sense, is spectatorial; acceptance is agential.

Put this way, the agential stance might seem to involve no more than adding a future-oriented desideratum to what the spectatorial stance delivers.

If so, then besides being worthy of belief—thereby satisfying the spectatorial requirements—a conclusion should also be a means to further epistemic achievements and should incorporate some provision for sidelining or downplaying trivial epistemic gains. This is not right, for it treats fruitfulness as a merely instrumental value. The agential epistemic stance insists that promoting further epistemic ends is intrinsic to acceptability. It underwrites our ability to exploit epistemic achievements and limitations to further our epistemic goals. It enables us to orient our inquiries so as to make achievements that can be leveraged.

There is, then, a pragmatic moment in epistemic acceptance. Deeming a candidate commitment worthy of acceptance involves holding that its incorporation will serve one's evolving epistemic ends. This does not make its acceptance merely instrumental. Nor does it make a consideration epistemically acceptable if its contribution is merely instrumental. Incorporating an untenable commitment into one's system simply to get a grant is epistemically impermissible, even if the work done under the grant contributes to understanding (see Elgin 2017:97; Firth 1981; Berker 2013). A commitment is acceptable only if its incorporation figures in the reflective equilibrium of a network of commitments. Part of what makes such a system acceptable, hence part of what makes its component commitments acceptable, is that the system constitutes a platform or preferably a springboard for further epistemic gains. Because a system in reflective equilibrium is, by design, something that can be extended and modified, the capacity for improvement is not an add-on. It is an integral to an acceptable network.

Autonomy, Community, Environment

Epistemic Ecology explicates the agential epistemic stance by interweaving three themes. The first considers conditions on epistemic agency. What does it take to be an effective epistemic agent—that is, someone who can act to foster her epistemic ends? It argues for an answer grounded in epistemic autonomy—the capacity of an agent to define and pursue her own epistemic goals. Such autonomy is an achievement, not a natural endowment. It involves thinking for oneself and taking responsibility for one's epistemic commitments. Social conditions have to be apt for epistemic autonomy to be both constituted and exercised. No agent will get far on her own. So the second theme concerns epistemic community. What sort of community underwrites, promotes, and

Epistemic Agency

benefits from epistemic agency? How is epistemic autonomy consonant with epistemic interdependence? How can members of such a community work together while respecting one another's epistemic autonomy? I suggest that an epistemic community is an epistemic analog of Kant's realm of ends. Its members make the rules and set the constraints that bind them because they collectively hold that being so bound will serve their epistemic ends. The third theme concerns the ways agents and communities engage with the domains they seek to understand. How do they represent the phenomena in those domains? How do they devise modes of access to those phenomena? I argue that individual and collective agency is in part constitutive of the objects of understanding. We structure domains—marking out the sorts of individuals and kinds that compose them, and the relations in which they can stand to one another—in order to make them accessible to us. We devise modes of representation and standards of interpretation that circumscribe the outputs we credit. Throughout, there is a concern with epistemic norms, standards, and criteria. Epistemology, as I construe it, is a normative discipline. Epistemic agents recognize that they are finite and fallible; in the exercise of their epistemic agency, they engage in critical reflective practices to overcome, evade, transcend, or exploit the impediments and obstacles to the understanding that they seek. That is, they constitute themselves as epistemic agents, their communities as communities of inquiry, and their domains as organized or organizable in ways that enable them to be understood.

It might seem that we can make short shrift of many of these issues by adopting a robustly realist stance. If the world (or anyway a large part of it) is as it is regardless of how we think it is, one might argue that our only epistemic challenge is to find out how it is. Regrettably, things are not so easy. Even if robust realism is true (which I strongly doubt and argue against in chapter 10), there remains the question of our access to the way things are, and there remains the question of how to tell whether we have succeeded. Rather than making things easier, realism may make them harder. Barry Stroud's work on skepticism underscores the legitimacy of the perennial worry that our best efforts leave a gap between evidence and truth. His point still holds if we replace a concern with methods with a concern for the reliability of mechanisms or the virtues of agents. In that gap, errors may arise (see Stroud 2018). The matter is further complicated because the way human epistemic agents have devised to access nature is via a variety of instruments, methods, measures, and techniques whose powers and limitations need to be understood. What justifies us in trusting our measuring devices? Why should we think that they accurately measure the magnitudes we are interested in? What justifies us in trusting our methods of inquiry—our experimental designs, sampling practices, proxies? What justifies our ampliative inferential strategies? Philosophers are familiar with worries about induction, but similar worries arise for statistical, analogical, and abductive inferences. These questions are not just about measurement, experimentation, and inference in general. They arise in real time in the real world when we consider how far to trust particular measures, experiments, and inferences. As we become epistemically sophisticated, our methods of inquiry become objects of inquiry.

Williams's conviction that the objects of inquiry are what is there anyway—that is, regardless of whether they are inquired into—is grounded in the recognition that the world is resistant to the will. Epistemically responsible agents cannot simply concoct alternative facts to suit their agendas. He is surely right about that. Nevertheless, epistemic agents are not passive onlookers. Hasok Chang (2004) shows how temperature emerged as a magnitude capable of being measured in tandem with and in response to efforts to design reliable, consistent thermometers. He does not, of course, contend that temperature was created ex nihilo. What he shows is that sensory predicates with vague and fluctuating boundaries were tamed and regimented into a sharp-edged, well-defined magnitude that could serve the purposes of science, industry, and commerce. The resistance of the world sharply constrained but did not wholly determine the parameters of the magnitude. Still, once it is settled that temperature is a particular magnitude and that a well-calibrated thermometer measures that magnitude, there is no room for subjective bias or wishful thinking. The resistance of temperature to my will is displayed in the fact that the temperature outside is below freezing today, despite the fact that I wish it were warmer. I will have more to say about this below. Although understanding must be responsive to the phenomena it concerns, realism about those phenomena does not determine the contours of that understanding.

One might worry that this approach sacrifices objectivity.[2] It emphasizes the epistemic ends and means of autonomous agents and well-ordered communities. What guarantee is there that we will not be led astray? Answer: there is no guarantee. That is why agential epistemology places so much emphasis on the capacity to identify and correct mistakes. Nevertheless, unbridled subjectivity is not a danger. A judgment or statement is subjective

Epistemic Agency

only if sincerity is determinative of correctness. A speaker could lie about, for example, whether she likes chocolate ice cream, but if she is sincere, her statement is acceptable. Nothing more is required. (Even this is not quite right, since it does not control for self-deception. But it is close enough that we can use it as a starting point, since it is plausible that self-deception about tastes in ice cream is rare.) Most judgments are not like this. Their correctness condition turns on the judgment being vouchsafed or capable of being vouchsafed by others. Initially, perhaps, our forebears didn't care who the others were or why they agreed. But pretty soon they realized that this much tolerance is too accommodating. As epistemic agents, we want agreement from those who are suitably competent to address the issue at hand. Thus, instead of mere intersubjective agreement, we insist on intersubjective agreement among those with appropriate expertise. We may still worry about the effects of peer pressure. So we buttress the requirement of expert agreement with a requirement that it be based on shared, reflectively endorsed reasons. We may go further and introduce measurement practices and instruments that take a particular judgment out of the hands of the experts and the reasons they can adduce (see Porter 1996). They still play a role in the background, since they provide the reasons to trust the instruments. At each step, we introduce constraints because we have learned what sorts of errors or infelicities we are vulnerable to and have identified ways to block or compensate for them. Relying on purely subjective reports in suitable cases—the liking or disliking of chocolate ice cream—is unobjectionable. We have no reason to think that a sincere speaker is wrong about her own taste. But in other cases, we worry that the judgment may be a product of idiosyncrasy, bias, or chance. In the quest for objectivity, we identify increasingly good reasons to think we have blocked idiosyncrasy, bias, and chance. Thus, a move that increases objectivity is justified not because it brings us closer to the truth, nor because we think it does, but rather because it pulls us away from practices that we have good reason to think are sources of error.

A fundamental question for epistemology is how we mold and regiment ourselves, our communities of inquiry, and our cognitive and material resources to promote and refine our epistemic ends. The agential stance is dynamic. We build on our achievements and learn from our mistakes. Indeed, we learn how to exploit our limitations to build on our achievements. Our predecessors did too. Epistemology should have something to say about how they leveraged their less than adequate accounts to make improvements.

I argue that autonomous epistemic agents constitute communities that collectively generate, sustain, justify, and improve understanding. *Epistemic Ecology* begins with an account of epistemic autonomy, construing it as a capacity for epistemic self-governance. It is a capacity to identify and foster one's own epistemic ends—to gain, extend, deepen, and refine knowledge and understanding. It asks what a finite, fallible actor must be like if she is to have and be able to exercise such a capacity. It considers the role education plays in fostering epistemic autonomy and fostering the appreciation of its value. Concerns about education are strangely absent from epistemology. Perhaps this is because much epistemology focuses on what I call the "end state"—the point where an agent either satisfies or fails to satisfy the conditions for epistemic success. But whether and how an agent satisfies those conditions typically depends on the resources he can draw on, and to a large extent they are a product of his education. We miss something crucial about our epistemic condition if we sideline the roles that education plays.

Recognizing that human beings are epistemically interdependent, I consider the nature, structure, and functions of effective epistemic communities. I argue that far from being antithetical to individual autonomy, a well-ordered epistemic community is supportive of it. Consensus is a good guide to acceptability only if the community consists of epistemically autonomous agents who can be counted on to voice their objections if they disagree. If members are inclined to agree in order to curry favor or preserve the status quo or stay out of trouble, their concurrence counts for little. If they are inclined to demur for reasons that have nothing to do with the merits of the case, their reservations are equally inert. Perhaps surprisingly, a well-ordered epistemic community requires that its members be autonomous. I go on to consider how epistemic agents—both individually and collectively—configure their domains, their resources, and their norms to foster their epistemic ends. This involves leveraging—building on current understanding, using current norms, standards, methods, and so on that are validated by the relevant epistemic community. That understanding must be flexible, admitting of correction and revision when the norms, standards, or methods seem inadequate and of expansion and deepening when opportunities arise. The goal is reflective equilibrium (see Elgin 2017:63–90). But since such an equilibrium is susceptible of being upset by further findings, which may reveal inadequacies or opportunities, the capacity to correct and extend is critical.

Epistemic Agency

Ecology is the study of the interactions of organisms and their environments. The interactions involve mutual adjustments as the organisms attempt both to fit into their environments and to mold the environments so as to make it possible for them to fit in. An ecological approach to epistemology studies the interactions between individual epistemic agents and their natural and social environments. Each adapts itself to accommodate the others. Epistemic agents form communities to foster their epistemic ends, and they adjust their ends to accommodate the constraints of nature and the requirements on epistemic community. Because they regard understanding as a good, they design practices, policies, and procedures to sustain and build on their achievements. Epistemic success, on this view, is not captured in individual "Eureka!" moments. It is a matter of sustainable achievements that afford a basis for further advances.[3]

It might seem strange to suggest that the world adjusts itself to the expectations of epistemic agents. But it was through the invention of calendars and clocks that time was reconfigured into repeatable units—years, months, hours, minutes. It was through the invention of taxonomies that the endless varieties of plants and animals were reconfigured into species, genera, and phyla. It was through the invention of moral categories that the demands we make of one another were reconfigured into networks of moral and legal rights and duties. In our quests for understanding, we reconfigure domains in ways that make them accessible. Once the reconfigurations are done, there are mind-independent facts about days, plants, and obligations. So despite the mutual adaptations, the world remains largely independent of the will.

Cognitive Control

Explications of understanding tend to bottom out, or at least plateau, in talk of grasping. An epistemic agent who understands why p grasps what makes it the case that p—for example, she understands why the house burned down when she grasps that the fire was caused by faulty wiring (Pritchard 2008:332). She understands p because q when she grasps the dependence relation between p and q. Thus, she understands that the milk spilled because the table was jostled when she grasps the relation between the table's being jostled and the milk's spilling (Grimm 2014:330). She understands a topic T when she grasps the way the elements of T hang together. Thus, an agent has some understanding of the tides when she grasps the way the moon's gravitational attraction gives rise to tidal behavior. She understands something of predator–prey relations when she grasps the pattern manifested in the Lotka–Volterra model, and a bit more when she grasps why the pattern holds (Elgin 2017:259–267).

The phenomena grasped may be material, mathematical, social, or institutional. They may involve norms. In understanding football, an agent may grasp, for example, the behavior of the players and referees, the rules of football, and the team's strategy for following those rules and exploiting the opportunities they provide. In understanding ethics, she may grasp the connection between lying and wrongdoing. She may further grasp why that connection is more complicated than it first appears (Hills 2016). Grasping, evidently, is the attitude an epistemic agent takes toward an item when she understands that item. What is that attitude?

It is not simply belief. An agent can believe something on the basis of testimony. Granted, she has to be able to interpret the testimony. (She would not believe it if she were told it in a language she did not understand.) But beyond that, little is required of her epistemically. She can believe, for example, that muon decays produce neutrinos, having been told as much by a source she trusts, but have no idea exactly what this involves. In that case, she does not grasp the relation between muons and neutrinos. Something more or different is required. It involves being able to properly manipulate the material one understands.

As used in epistemology, "grasp" is a metaphor. The term literally applies to a particular way of holding something in one's hand. We grasp a hammer, a tennis racquet, a door handle. Hands standardly do not figure in epistemic grasping. Epistemic grasping is not literal grasping. Still, to label the epistemological use metaphorical is not to disparage it. Nor is it to assume that the term's status as a metaphor makes it especially unclear. To call it a metaphor is to acknowledge that the epistemological application both depends on and diverges from our prior literal use of the term. I have argued that a metaphor functions by highlighting certain features of its literal referent and projecting them onto its new referent (Elgin 1988:59–70; 1996:197–204). Our fluency with the literal term guides our interpretation of the metaphor. That being so, it may pay to look first at the literal use of the term "grasp."

In properly grasping a tennis racquet, the player holds the racquet firmly but not stiffly in her hand. Her grip must be flexible enough that she can

Epistemic Agency

bend and twist her wrist to manipulate the racquet dexterously, enabling her to control the ball. Literal grasping is agential. It is something people do—typically, something they do on purpose. (You take the racquet in your dominant hand, hold it firmly yet flexibly with the V between your thumb and your forefinger aligned with the axis of the racquet.) Grasping thus differs from seeing—another term that is commonly used metaphorically in epistemological contexts. Often, to literally see something is simply to have it register in your visual field. And to metaphorically see something is simply to be aware of it. The lecturer sees that the audience has lost the thread of the argument. She doesn't have to do anything special to discover this; their bewilderment is all too apparent. The commuter sees that the train is late. He simply senses that, although he arrived at the usual hour, he's been waiting for an unduly long time. To literally see , you needn't do much beyond being visually aware of your surroundings. To metaphorically see is to relax the restriction to the visual; it is a matter of awareness. Seeing is thus more passive than grasping. If understanding involves metaphorical grasping, understanding is agential. It is a matter of doing or of being in a position to do something.

Allison Hills maintains that intellectual grasping is a matter of cognitive control. I think she is right. But the control is a more comprehensive competence than she suggests. She focuses on understanding why, saying that someone who grasps that p because q (where it is true that p because q) understands why p. "When you grasp a relationship between two propositions, you have that relationship under your control. You can manipulate it" (2016:663). Strictly, this cannot be correct: p and q are propositions. They stand in some relation to each other. That relation is what it is, regardless of anything the agent does or can do. If "dinosaurs are extinct because a giant meteor hit Earth" is true, then someone who grasps the connection between "dinosaurs are extinct" and "a giant meteor hit Earth" understands why dinosaurs are extinct. Even so, he has no control over the relation between these facts or the propositions that express them. He cannot manipulate them. Still, there is something right about Hills's position.

In characterizing the sort of control Hills ascribes to someone who understands why p, she says:

If you understand why p (and q is why p) then . . . in the right sort of circumstances you can successfully

    (i) follow some explanation of why p given by someone else.

    (ii) explain why p in your own words.

(iii) draw the conclusion that p (or that probably p) from the information that q.

(iv) draw the conclusion that p' (or that probably p') from the information that q' (where p' and q' are similar but not identical to p and q).

(v) given the information that p, give the right explanation, q.

(vi) given the information that p', give the right explanation, q'.

(2016:663).

The abilities described in (i)–(vi) do not merely supply evidence; they are not instrumental. They are, according to Hills, at least partly constitutive of understanding why. To understand why p is to be able to do something involving explanation and inference by drawing on one's relevant epistemic commitments. Although you cannot manipulate the relation between p and q, you can manipulate the information about that relation in generating explanations and making inferences. Moreover, if you understand why p, you are evidently supposed to know how to do all of (i)–(vi).

The competences on the list fall into two groups. Those in (i), (ii), (v), and (vi) are abilities to give or appreciate explanations. Those in (iii) and (iv) involve inference. I will argue that the requisite abilities are broader than Hills recognizes. Explanation is but one way of manifesting understanding— a way that requires a measure, often a significant measure, of linguistic ability. Inferences that are neither deductive nor probabilistic can manifest understanding.[4]

Even if we restrict our attention to understanding-why, there are serious limitations to Hills's characterization. To be able to explain requires articulateness. A young child, having seen the dog knock over the milk carton, may understand why the milk spilled, even if he cannot explain the event. An inarticulate auto mechanic may understand why the check-engine light is malfunctioning. He can readily fix the problem, and his doing so manifests his being able to distinguish among a variety of potential causes and cures of the malfunction. But he cannot explain what caused it or why his remedy works. A ballet master may understand why a dancer's tendon is strained, even though what he appreciates about her stance is too fine-grained to put into words. In each of these cases, the protagonist may be able to display his understanding, even though he cannot verbalize it. The child tips the milk carton to show what the dog did; the mechanic confidently rewires the electrical connection to the check-engine indicator, curing its tendency to come on arbitrarily; the ballet master takes his hand and subtly adjusts the dancer's stance so that her tendon is more relaxed. All three may be incapable not

Epistemic Agency

only of explaining but also of following a verbal explanation. Words are not their medium. But all are adept at giving and following a relevant demonstration. Privileging the verbal makes understanding why overly intellectual.[5]

Nor is the requisite connection between following an explanation for why p and giving an explanation in your own words as straightforward as it might seem. The need for a requirement like (ii) is clear. Merely parroting an explanation why p does not constitute understanding why p. But how, and how far one's own words must diverge from the original explanation is not obvious. Some ways of putting a passage into one's own words amount to no more than a trivial paraphrase. They are not quite parroting but are not far from it. If a student were to paraphrase her professor's explanation of why water expands when it freezes by transforming the active voice into the passive voice, replacing sentences with obvious logical equivalents, and substituting synonyms for synonyms, it would not constitute or be evidence of understanding why water expands when it freezes. If a student with a limited vocabulary in the language of instruction could not put the teacher's explanation in any words other than those that were trivially equivalent to the original explanation, this would not by itself seem to constitute a defect in his understanding of the phenomenon. Inherent in Hills's demands involving explanation are implausible requirements of articulateness and fluency.

The capacity to draw conclusions also deserves scrutiny. Given the information that q, the understander is supposed to be able to draw the conclusion that p. (We assume that the information is cast in terms that are intelligible to the agent.) But what sort of inference is involved in drawing that conclusion? Hills does not say. It seems she thinks the inference is deductive or probabilistic. But anyone reasonably adept at logic can infer a host of deductive and probabilistic consequences from any proposition whatsoever. He doesn't even have to understand the nouns or verbs. I'll suggest below that the ability to make inferences is important, but the range of reasoning that counts as inferential must be broadened.

To prevent parroting from qualifying as a display of understanding, Grimm says that the epistemic agent's grasp of the relation between p and q should sustain counterfactuals (2006:532–533). That is, the agent should be able to say not just what the relation between p and q actually is but also what it would have been in suitably similar counterfactual situations. This may be plausible for contingent matters, even those where the "because" is underwritten by scientific laws. The agent grasps the connection between the rock hitting the window and the window shattering when she appreciates that, ceteris paribus, if the rock had not hit the window, the window would not have shattered, at least not in the way that it did. But a counterfactual account cannot accommodate mathematical understanding, since the connections between mathematical propositions are necessary. To understand why Mother cannot evenly divide twenty-three strawberries among her three children without dividing any strawberry, it makes no sense to ask what would have been the case if twenty-three had been divisible by three (contra Lange 2017:6). Mathematical relations are necessary; they hold in all possible worlds. Hills's account fares better. Rather than asking what would have been the case had the situation vis-à-vis p been different, she asks what would be the case for a distinct but relevantly similar p'. She can then anchor the agent's understanding of the relation between p and q in what would be the case (for sharing out strawberries) if Mother had twenty-four strawberries instead. The recognition that to understand that q is why p requires an appreciation of the relation between various values of p' and q' is a recognition that understanding why involves locating p and q in a neighborhood of a logical space—a space of relevant alternatives to p and q.

It is important that p' and q' are like p and q in relevant respects. This is implicit in Hills's discussion and is not, on the face of it, problematic. But it is worth noting that the agent presumably has to be able to tell which values of p' and q' are appropriate. There are any number of ways that the fact that the rock hit the window might be described. Some of them would be unintelligible to the agent; some, although intelligible, would not lead her to give the information that the window shattered. (An easy example: "Billy did what he was explicitly told not to do"—viz., throw a rock at the window. This will not underwrite an inference to "The window shattered." It might underwrite one to "He's going to be in big trouble.")

The characterization that looks, on the face of it, to be a characterization merely of understanding why—a relation between two propositions—thus implicates a broader understanding of a subject. To satisfy Hills's requirements, the epistemic agent has to draw on an understanding of the epistemic environment in which p and q (and their ilk) are located. Without that broader understanding, the agent would be in no position to satisfy (i)–(vi).

The emphasis on cognitive control is correct. It consists in knowing how to exploit the available information and other epistemic resources to promote

Epistemic Agency 23

one's epistemic ends. Hills considers the know-how characterized in (i)–(vi) to be necessary for understanding why. I disagree. The focus on explanation and deductive or probabilistic inference strikes me as too narrow. Rather, I suggest, that the conditions that Hills lists characterize a particular realization (or a constellation of particular realizations) of a broader cognitive ability. To grasp *p* or *why p* or *that p* is to be able to effectively wield one's information pertaining to *p*. It is to be able to exploit the information to further one's cognitive and practical ends. If an agent can explain why *p*, she displays some understanding of *p*. If she can handle variants on *p*, explaining why if *q′* were so then *p′* would be so, she displays greater understanding. (She can demonstrate that she is not parroting.) If, given the information that *q*, she can deduce that *p*, she shows greater understanding. But does understanding why require such verbal dexterity? Or are these abilities just some ways of manifesting her understanding? Earlier, I objected that Hills's account seems to place too great a value on being articulate. I suggested that an agent could display understanding without being able to explain. The child, the auto mechanic, and the ballet master evidently have what is for them inarticulable understanding.

Peter Lipton adduces several examples. One concerns displaying an understanding of the retrograde motion of the outer planets (2009:45). Someone who cannot explain why Mars seems to reverse its direction might manifest her understanding by properly manipulating an orrery. She moves the models of the planets in a way that shows why Earth's moving more quickly than Mars gives rise to the perception from Earth as of Mars reversing its direction. This allows for parallels to Hills's conditions, replacing explanation with demonstration. The agent can

(i′) follow a demonstration of why *p* given by someone else: she can, for example, follow a demonstration of the retrograde motion of Mars given by someone using an orrery;

(ii′) give her own demonstration of why *p*: she can manipulate an orrery, or another model of the solar system with movable parts, to show why Earth's moving faster than Mars is what gives rise to the phenomenon;

(v′) given the information that *p*, provide a demonstration based on *q*

Given the information that Earth's year is shorter than Mars's year, she could use the orrery to show why, from the perspective of an observer on Earth, Mars appears to reverse its direction;

(vi′) given the information that *p′*, provide a demonstration based on *q′*.

Given the information that Mars's year is shorter than Jupiter's, she could use the orrery to show why, from the perspective of someone on Mars, Jupiter would appear to reverse its direction. Alternatively, she could invent imaginary planets and show how, from the perspective of one, the other would display retrograde motion. Understanding why does not always involve an ability to explain; it often involves an ability to demonstrate or illustrate. In manipulating the orrery to demonstrate retrograde motion, the student's actions exemplify the relevant dependence relation. Instead of saying why Mars appears to reverse its direction, she shows why. If she can embed such showing in a suitable range of counterfactual circumstances, her epistemic accomplishment seems at least as robust as that of the student who can put the relation into words.

Conditions (iii) and (iv) involve drawing conclusions—that p (or that probably p) from the information that q, and that p' (or that probably p') from the information that q'. But the ability to engage in deductive and probabilistic inferences about p by itself manifests little if any understanding of why p. (It may, of course, manifest considerable understanding of deductive and probabilistic reasoning.) I suggest that understanding involves a broader range of inferential capacities. In addition to deductive and probabilistic inferences, grasping the connection between p and q might involve abductive inferences (Lipton 2004), analogical inferences (Nersessian 2008), perhaps implicatures (Grice 1989), maybe explications. The agent might, for example, grasp that the best explanation of why p is that q. By inferring that the best explanation of the footprints on the beach is the recent presence of a walker on the beach, Robinson Crusoe understands why there are footprints. The agent might recognize that the behavior of predatory lenders is analogous to the behavior of sharks, thereby understanding why impoverished people, like small fish, are especially vulnerable. The agent might appeal to implicature to grasp why "There is a service station around the corner" can be a relevant and appropriate reply to "Do you have jumper cables?" The agent's grasp of a complicated phenomenon (perhaps the life cycle of an organism or the course of a disease) might involve recognizing that although her model captures important features, it is not accurate in every case and does not reflect every causally relevant aspect.

The ability to make a broad range of relevant ampliative inferences and to refrain from making irrelevant ones relies on a grasp of the cognitive context. So propositional understanding presupposes objectual understanding.

One cannot understand why the train is late or why Mars displays retrograde motion without having a more comprehensive understanding of a relevant range of phenomena—contemporary public transportation, on the one hand, and planetary motion, on the other. And one cannot have such a comprehensive grasp without relying on the resources supplied by a relevant epistemic community.

Understanding, I maintain, is agential. Thinkers are active, not passive. Their perspective is first personal, not third personal. They need to situate themselves in epistemic contexts, understand themselves as epistemic agents who have alternatives to choose among, resources to draw on, projects they want to promote. Agents, it seems, must be autonomous.

2 Epistemic Autonomy

The Tension

Autonomy and interdependence seem antithetical. An autonomous agent thinks for himself and acts as he sees fit. He is, according to a prevailing stereotype, a rugged individualist. Nobody tells *him* what to think or what to do! Interdependent agents rely on one another. They regularly confer, eventually achieve consensus, then jointly act on that consensus. They correct, refine, and extend one another's thinking. On the basis of shared commitments, they work together to achieve common goals.

Although much of the philosophical discussion of autonomy occurs in ethics and action theory, the tension between autonomy and interdependence also arises in epistemology. An individualist stance frames the epistemological problematic we inherited from Descartes (see Code 1991). In its quest to secure knowledge, the Cartesian ego fends for itself. It purposely, even ruthlessly, distances itself from other minds. But if autonomy requires individualism, we are in a bind, for as has become increasingly evident, epistemic agents are ineluctably interdependent. We cannot survive on our own (see Grasswick 2018).

Jettisoning autonomy is not an option though. Individual thinkers are rightly blamed for sloppy reasoning, jumping to conclusions, ignoring base rates, and a host of other epistemic sins and indiscretions. They are rightly praised for rigorous arguments, creative ideas, fruitful conceptual innovations. Such praise and blame are warranted only if their cognitive condition is under their control. To be praiseworthy or blameworthy, they must be responsible. An agent is responsible only for what he freely does or freely refrains from doing. He is not responsible for things he is powerless to affect.

Responsibility requires autonomy. Is there a way to reconcile epistemic autonomy with interdependence? I will argue that there is. Rather than being antithetical, I will urge, epistemic autonomy and interdependence are mutually reinforcing.

Individualism

By tradition, epistemology is individualistic. The isolated thinker is supposed to be solely and wholly responsible for her epistemic condition. Relying only on her personal resources, she either knows or does not know. Descartes was the champion of rugged epistemic individualism. The Cartesian ego credited only deliverances that it was, on its own, at that very instant, in a position to validate. Not only testimony but also long-term memory, perception, even extended inferences were initially rejected because their deliverances might, for all the ego can immediately tell, be wrong. The hope was to ground all genuine knowledge on a base of convictions that an epistemically isolated ego could not doubt. Unfortunately, by the end of *Meditation II*, the game was up (Descartes 1979). Since the ego could not prove that God exists and is not a deceiver, it ended up knowing very little. Skepticism being an unattractive option, less rugged forms of individualism emerged.

A certain amount of scaffolding was needed to get beyond the Cartesian framework. Not all putatively perceptual deliverances, apparent memories, outputs of introspection, or seemingly plausible inferences are epistemically creditable. A faint glimpse out of the corner of one's eye, a hasty mental calculation while emerging from anesthesia, a dim recall of a trivial event (such as the third-place finisher in the fifth-grade sack race), the heartfelt conviction that one is in love with a person one has never met—none of these is creditable. Conditions on epistemic acceptability had to be framed to exclude such deliverances. Still, many mundane deliverances were held to pass muster. The sighting of a not unexpected, middle-sized object in good light in the center of one's visual field; a careful, double-checked, not too difficult calculation; a clear memory of a significant, not implausible event came to be acknowledged to be pro tanto creditable.

Deliverances of perception, introspection, memory, and calculation answer to conditions that an agent can satisfy on her own. They are, let us say, individualistically grounded. If individualistically grounded deliverances exhausted our epistemic resources, it would be straightforward to

Epistemic Autonomy

construe thinkers as epistemically autonomous. Each would be solely and wholly responsible for her beliefs' meeting or failing to meet the conditions on epistemic acceptability.

Interdependence

But individualistically grounded factors do not come close to exhausting our epistemic resources, satisfying our epistemic needs, or accounting for our epistemic successes. We depend heavily on one another to supply information we do not ourselves possess and could not—or not easily—get on our own. A vast number of our beliefs are based on testimony: the shortest route to Porter Square, the date of the Norman Conquest, the atomic number of gold, the departure time of the train to Bedford. The problem is that testimony seems inherently precarious or at least considerably more precarious than other grounds. Informants are fallible; they are prey to the same vulnerabilities in perception, memory, introspective acuity, and inference as we are. In accepting testimony, we apparently inherit our informants' vulnerabilities on top of our own. Informants may, moreover, be careless, misinformed, or dishonest. As the length of a testimonial chain grows, the risk increases. And even if each person is a good judge of whether she herself is being scrupulous, we typically have no way to check on the moral and epistemic bona fides of those whose word we rely on. Often, we do not even know who they are.

It might seem that we could easily diminish our reliance on informants. Rather than asking a passerby for directions, we could download a map. Instead of checking with a lab partner, we could consult the periodic table of elements. Rather than querying the ticket agent, we could look at the timetable. And so forth. Although the documents we draw on may be more reliable than an arbitrarily chosen informant, they do not eliminate our dependence on others, for such checking just substitutes one sort of testimony for another. Someone made the map, devised the table of elements, wrote the timetable. In trusting the documents, we implicitly trust those who made and vetted them.

Besides relying on others for information, we depend on them for instruction. If you know how to calibrate an instrument, calculate an integral, tune a piano, judge a dive, it is likely that someone taught you. The knowledge you glean from such activities is derivative from that teaching. If your instructor was incompetent or slipshod, you might, through no fault of your own, regularly and systematically misinterpret the resources you appeal to and thus misjudge. Nevertheless, one might think that reliance on instruction is not so serious a threat to autonomy as our dependence on testimony is, for once a subject is suitably trained, she can conduct the requisite investigation for herself. She can, in Wittgenstein's terms, kick the ladder away once she has climbed it (1947:§6.54). This is not quite right though. The current user did not design the measuring device or figure out how to calibrate it (Chang 2004). The current judge did not set the standards for assessing a dive or specify the respects in which and degrees to which deviating from those standards constitutes a flaw (USA Diving 2019). The current student did not construct the table of logarithms (Klein 1972). The epistemic standing of the beliefs an agent forms by deploying such techniques and devices depends on the adequacy of the techniques and devices themselves. That depends on how they were designed, created, and validated. If a measuring device is poorly calibrated, if the table of logarithms is inconsistent, if the standards for assessing a dive are jointly unsatisfiable, the results will be unreliable.

I've chosen examples that are well entrenched. Others have used them effectively in the past. This is evidence of their adequacy. The examples are collective achievements. Even if a single individual invented a particular technique or device, in order for it to be epistemically estimable, others had to validate it. Users typically have no idea what justifies many of the techniques and devices they rely on. They have no clue, for example, how logarithms are generated or how gas gauges are calibrated. People who insist "Nobody tells *me* what to think!" are wrong. That's their good fortune.

Human beings are epistemically interdependent. It would be ludicrous to contend that an agent does not know that her gas tank is nearly empty or that the atomic number of gold is seventy-nine or that π is an irrational number simply because she did not figure it out for herself. If epistemic autonomy is incompatible with interdependence, we are not autonomous. But the objections I have raised are directed against epistemic individualism, not against epistemic autonomy. We go too fast in identifying the two.

Action

So far, I have construed autonomy loosely as freedom. What sort of freedom is at issue here? To be free in the relevant sense is not to be utterly unconstrained, bouncing about like a gas particle in random motion. Nor is it to

Epistemic Autonomy

be uninfluenced by others. Personal autonomy is a matter of individual self-governance. There is an analogy to the political realm. An independent nation is autonomous in that it is self-governing. It makes and enforces its own laws. No other nation has dominion over it.

On a standard model of action, an actor desires p and believes that he will gain or improve his prospects of gaining p if he does q (see Davidson 1980). His desire for p thus drives him to do q. It might seem that because he is driven by his desires, an actor does not deliberate. He is some sort of stimulus–response machine. As soon as a desire presents itself, he acts to satisfy it. Some acts and some actors may fit this profile. It is plausible to think that some nonhuman animals do so. On seeing a mouse, an owl desires it and immediately strikes. But not all animals—indeed, not all predatory animals—are like that. A clever cat holds back. Rather than immediately pouncing, she bides her time. She waits until the mouse is far enough from its hole that it cannot escape. Then she pounces. Something akin to deliberation seems involved. Other animals are even more sophisticated. Prides of lions work together to stalk and catch their prey. They apparently coordinate their behavior, learning from experience what strategies are effective. They teach their cubs how to hunt—that is, how to work together to catch prey. They may be driven to act by their desire for the gazelle. But even if their behavior is instinctive, it is evidently not impulsive.

Like the impulsive owl, humans sometimes act unthinkingly to get what they desire. The marshmallow appears on his plate, and Max immediately, unthinkingly, pops it into his mouth. But much human behavior is more complex. Human beings have conflicting desires. Sometimes, we are aware of the conflict. In that case, let us say that we have *manifestly conflicting desires*—desires that the desirer recognizes cannot all be satisfied. Confronted with such desires, we can deliberate about which is strongest. Considerable thought may go into weighing alternatives. It may involve assessing not just the attractiveness of our conflicting desiderata but also the probability of achieving them. Let us call the product of such deliberation the actor's strongest desire. A deliberative actor is not impulsive, but his action is not guided by any evaluation of final ends. His final ends are fixed by whatever he happens to take himself, on balance, to most strongly desire. They are given rather than chosen. Such an actor, human or not, thus seems to be a victim of circumstances. He finds himself with certain desires, perhaps thinks about which desires are strongest, perhaps deliberates about how to achieve his desires. But he is bound to attempt to achieve whatever he takes to be his strongest desire.

Although he may be able to figure out which desire is strongest, he cannot determine whether that desire, or indeed any of his desires, is worthy of satisfaction.—So the standard model suggests. It might seem that from a subjective perspective, this is no problem. Given that he desires p more strongly than he desires any competing alternative, he deems p worthy of satisfaction. No other desire overrides it. Harry Frankfurt's unwilling addict belies this (1971:12). Although the unwilling addict strongly desires the drug she is addicted to and can't help desiring it, she deeply regrets her desire. She cannot overcome her addiction, but she desires to not desire the drug. She is overpowered by the desires she finds herself having. They dictate her behavior.

Belief

Discussions of action typically focus on desire. The actor is in an important sense unfree because his actions are determined by desires he simply finds himself with. But action is the joint product of belief and desire (see Davidson 1980). So an actor is equally unfree if he is driven to act on the basis of beliefs he simply finds himself with. If there are such beliefs, this becomes an issue for epistemology.

Phobias seem to fit the bill. They are irrational fears with specific targets. Aerophobia is an irrational fear of flying; arachnophobia, an irrational fear of spiders. Fear is an emotion though. So it might seem that phobias are irrelevant to our problem. But emotions are not just spontaneous upwellings of feeling. They embed beliefs (see Elgin 1996:147–149). To fear something involves believing that it is dangerous. Aerophobia thus embeds the belief that flying is dangerous. If one cannot help but believe that flying is dangerous, one's seemingly unreasonable avoidance of flying is driven by an inescapable belief. Indeed, if one believes that flying is dangerous, there is nothing unreasonable or irrational about desiring not to fly.

Besides diagnosable phobias, entrenched stereotypes, deep-seated prejudices, and products of wishful and fearful thinking are plausible candidates. Despite her dismal prognosis, optimist Olga convinces herself that she will recover. She willfully blinds herself to the extent to which her test results indicate otherwise. Despite no-hidden-variable theorems, determinist Dan believes that the Heisenberg uncertainty principle is false; the grip of the causal law on his reasoning is too tight for him to escape. Such attitudes are largely impervious to evidence. When someone is under the sway of such an attitude, it seems to make sense to say that he can't help but believe what he does. If this is a literal description rather than a rhetorical trope, such a person is doxastically fettered. Whether or not a doxastically fettered actor is blameworthy, his state is epistemically defective.[1]

Phobias, prejudices, inaccurate but entrenched stereotypes, and products of self-deception are failures of rationality. Although they are serious impediments, it may seem that they have little bearing on epistemology. But the locution "*x* can't help but believe what she does" seems equally true of ordinary believers vis-à-vis their ordinary beliefs. If Brenda sees a blue car in broad daylight in the center of her visual field and has no reason to think that she is perceptually impaired or that the setting is abnormal, she can't help but believe that there is a blue car in front of her.

Many of our beliefs are products of our culture, our education, our encounters with motley crews of informants. Many were unsought. Many were unvetted. Some may be unwelcome. We just find ourselves with them, having been brought up in a particular society and having had a certain range of encounters. They are not individualistically warranted and, for all we know, may not be warranted at all. Inasmuch as we live in an information-rich and largely unfiltered environment, much testimony is likely to, in Plato's words, "influence us unawares" (1974, *Republic* III). We have no idea how we came to have or what would justify some of the views that are common currency in our society. Moreover, our experiences are parochial. Beliefs formed on the basis of them may well be skewed. Our education is spotty. We may know how to take square roots, but have no idea why the method works. We may know the date of the signing of the Magna Carta, but have no clue how that date was established. To be sure, some beliefs that are responsive to evidence and some products of testimony are well founded. The agent can't help but believe them, given that support. Still, no more than the suspect beliefs are they up to her. They are not products of free choice.

A cat might have a first-order belief about where a mouse is. Wanting the mouse, she might act on that belief. But the cat is not responsible for the belief any more than she is responsible for the desire. Both arose unprompted, stimulated by that distinctive, mouthwatering mouse aroma. She cannot deliberate about whether that smell is actually a good way of telling whether there is a mouse behind the bookcase. If, because of its etiology, a person's belief is not under his control, then like the cat, he is not responsible for it. Beliefs just happen to him. This creates a problem.

The Aim of Belief

Belief aims at truth. To find oneself believing that p is to find oneself taking p to be true. On learning that p is false, one deems believing p to be a mistake. A believer, as such, wants to believe only what is true. This is so, regardless of the etiology of belief. But just as we can have manifestly conflicting desires, we can have manifestly conflicting beliefs. Ned finds himself believing that p and believing that q, although he recognizes that p and q conflict; p and q cannot both be true. Perhaps he believes that the ethics seminar meets on Monday and believes that it meets immediately after the logic lecture which meets, not on Monday, but on Tuesday. Something is amiss.

Taking a cue from the way he handles manifestly conflicting desires, he might ask himself which one he believes most strongly. He can answer that. He believes, let us say, p more strongly than he believes q. The issue here is how confident he is that p. It is not what subjective probability or credence he attaches to p. In the absence of a reason to think that the strength of his belief correlates with the likelihood of its being true, he has no grounds for thinking that preserving the stronger belief and abandoning the weaker will promote the aim of believing only what is true. To deliberate in a way that promotes that aim, he must critically reflect on the supports for his beliefs. To figure out which of the conflicting beliefs he ought to retain, he needs to assess the two beliefs and decide whether they and their supports stand up to scrutiny.

He thus must adopt a second-order stance. But if his second-order stance is one in which he just finds himself harboring beliefs about his various first-order beliefs, he is still in a bind, for the problem will simply recur. Should he discover a conflict between his second-order beliefs or between a second-order belief and a first-order one, he will have to adopt a third-order stance to determine which he believes most strongly. A regress threatens. It won't go far. Before long, he is apt to have no views about which of his beliefs about his beliefs about his beliefs . . . is stronger. Moving up the hierarchy is thus not likely to help. Nor is that the only problem. If he finds himself with a higher-order belief that casts doubt on a lower-order one, he still has no basis for preferring one to the other. They are just two beliefs that he finds himself

Epistemic Autonomy

with—one, curiously enough, being about the other. Even if his higher-order beliefs ratify his level of confidence in his original belief, he is not out of the woods, for although he more strongly believes p and more strongly believes that he more strongly believes p, and more strongly believes that he more strongly believes . . . , this gives him little reason to think that retaining p and abandoning q promotes his goal of believing only what is true. If he is subject to confirmation bias, this self-reinforcing, ill-advised hierarchy of beliefs is exactly what one would expect.

Merely adopting a higher-order stance is not enough. To resolve the difficulty, Ned must adopt a critical stance—one from which he can assess his beliefs. He needs to be able to identify and evaluate his reasons—the considerations he takes to bear on whether p is true. Distorting factors such as salience bias, proximity bias, and confirmation bias strengthen beliefs illicitly. He thus needs to be able to exclude irrelevant influences from his thinking.

His stance must be more than merely a standpoint from which to find faults. A sportscaster doing a play-by-play broadcast describes the events on the field, identifies good and bad moves, characterizes the strengths and weaknesses of the various players, discerns the teams' strategies and tactics, contextualizes the current game in the season and the history of the rivalry, and so on. Still, his perspective is third personal. Although he recognizes opportunities, he cannot exploit them; although he discerns flaws, he cannot correct them. He embodies and conveys a rich understanding of the game as it unfolds, but in his current role, he is an onlooker. His critical stance is spectatorial. If Ned is just a spectatorial critic of his own mental life, he is not in a position to promote his truth-seeking goal. He can only recognize with regret how he falls short. What he needs is a way to use his critical stance to improve his epistemic lot. He has to be able to correct the flaws. To do that, his critical stance must be agential. Earlier, I characterized beliefs as largely involuntary. Given our circumstances, we cannot help but believe what we do. So it might seem that Ned's predicament is irresolvable. Not quite.

From Belief to Acceptance

To address the difficulty, let us distinguish between belief and acceptance. In articulating the distinction, I draw heavily on the work of L. Jonathan Cohen (1992). As I read it, Cohen's distinction is an explication of the commonsense notion of belief. "Believing that p" has two components. One is feeling or being disposed to feel that *p* is so. This is largely involuntary. Cohen reserves the term "belief" for this component. The other is acceptance. To *accept* that *p* is to be willing to use *p* as a premise in inference and as a basis for action when one's ends are cognitive. One's ends are cognitive when one attempts to know, understand, comprehend, or otherwise grasp how things stand in a given domain.

Although I draw heavily on Cohen, my conception of acceptance is both broader and narrower than his. It holds that an agent must not only be not only willing but also able to use *p* in inference and action. Moreover, I eliminate "as a premise" because, on my view, we also accept rules, standards, and methods when our ends are cognitive. In my terms, to be willing to use modus ponens in cognitively serious inferences is to accept it. My conception is narrower than Cohen's because the only inferences it applies to are assertoric inferences. I set aside accepting for reductio, since it is too short-lived for my purposes (see Elgin 2017:19).

To be willing to use something as a basis for inference or action is to be willing to deploy it to generate results that, because they are so generated, are pro tanto worthy of one's reflective endorsement. That being so, experimental methods and other investigative techniques qualify for acceptance when the agent has reason to think that they yield results that promote her epistemic ends. Similarly, epistemic standards—such as standards of evidence, relevance, and accuracy—qualify for epistemic acceptance when the agent has reason to think that satisfying them promotes her cognitive ends.

An autonomous epistemic agent's reflective endorsement—be it of a proposition, a principle, a norm, or a method—is, I suggest, a matter of acceptance rather than belief. Acceptance is voluntary. The agent therefore can exercise epistemic self-discipline; she can accept or reject considerations as she sees fit (see Weatherson 2008). Someone who, because of her phobia, cannot help but feel—that is, in Cohen's sense, believe—that flying is dangerous may nevertheless rely on the relevant research and accept that flying is safe. She is then willing (if, she admits, unreasonably reluctant) to fly, willing to advise others to fly, and willing to use "flying is a safe mode of travel" as a premise in her inferences.

Acceptance neither requires nor always eventuates in belief. A scientist might accept *h* as a working hypothesis, and devote years to teasing out its consequences, without ever quite believing that it is true. Perhaps it is

Epistemic Autonomy

a promising hypothesis for which he never is able to garner sufficient evidence. In that case, although it may be well supported, it remains somewhat speculative. Perhaps he accepts it without specific reservations but expects that in the long run it, like all current science, will be superseded. Yet a third possibility is that he accepts it as an idealized model that does not even purport to be true. If so, regardless of the support, it is not a candidate for belief (see Elgin 2017:249–271).

The cat's propensity to act on whatever she believes is not a problem for her. Even if she harbors conflicting beliefs, she is unaware of the conflict. But because we humans recognize that our beliefs can conflict and recognize that belief aims at truth, our situation is different. Nor is this just a problem that emerges with logically inconsistent beliefs. The real threat is not inconsistency but non-cotenability. Belief contents that are not logically at odds may fail to be cotenable. Meg may believe that she needs to devote several hours of focused attention to finish the project that is due tomorrow, and believe that she has plenty of time her to go out for drinks with friends, postponing work on the project until she gets back. Either belief might be plausible, but their conjunction is not promising. If p sufficiently lowers the probability that q, or sufficiently diminishes the plausibility of q, an agent who finds herself believing both p and q ought not accept both. In light of one, the other does not reach the threshold of acceptability. The problem ramifies. We cannot resolve it simply by assessing beliefs pairwise. We need to evaluate how tenable entire networks of belief are. Individual commitments are acceptable only if they belong to systems of mutually supportive commitments—commitments that are reasonable in light of one another.

Because belief aims at truth and Ned has (or might easily have) conflicting beliefs, if he is to promote his goal as a believer, he must be in a position to endorse or repudiate some of his beliefs. He must be able to do so on the basis of considerations that he reasonably takes to advance his truth-seeking agenda. And he must be able to act on his decision—not merely applaud or rue his findings. Even if he can't help but believe what he does, he can accept or reject considerations, depending on whether he takes them to promote his cognitive objectives, and he can live his life accordingly. So he is an agent. As such, he can be held responsible for the way he conducts his cognitive life.

This raises the question of how he ought to conduct his cognitive life. Ned ought to accept only considerations he reflectively endorses, for these are the only ones he has reason to think his epistemic principles vindicate. This is not to say that he does or should examine candidate commitments one by one and decide, after due reflection, which of them are worthy of acceptance. He develops dispositions to treat wide swaths of considerations as default acceptable. A putative sighting of a car on Main Street in Middle America is default acceptable. Because he recognizes that such a car is the sort of thing one would expect to see, its acceptance raises no red flags. But he is sensitive to anomalies. When a consideration fails to fit with what he default accepts, he suspends judgment until the challenge is answered. Although he is disposed to accept deliverances of well-illuminated middle-sized objects at the center of his visual field, he refrains from default accepting a putative sighting as of a camel on Main Street. Eventually, he may come to accept it, but its presence on the scene needs to be accounted for before he is willing to credit it.

The worry is that this just pushes the problem up a level, for the issue of acceptability arises for principles, both tacit and explicit, as well. This is so, but it does not lead to an infinite regress. Since Ned is an agent, he can start and end a justificatory chain as he sees fit.[2] He can, that is, decide, on reflection, that the type and degree of justification he has are sufficient for his cognitive purposes. This may, but need not, lead him to seek justification for his second-order principles, but it need not (and should not) lead him to relentlessly demand ever higher orders of justification. He is responsible for his commitments because he takes responsibility for them. At whatever level he stops, he concludes that the considerations he accepts provide a good basis for inference and action in light of his current cognitive ends.

Whether driven by evidence and argument or by irrational forces, Ned cannot help but believe what he does. He is, to use a Kantian term, doxastically heteronomous. Perhaps surprisingly, however, he is *epistemically* autonomous. His autonomy flows from his voluntary acceptance of only considerations that he reflectively endorses. Forming an opinion by deciding whether a consideration is worthy of reflective endorsement is something he does, not just something that happens to him. Worries about provenance pose no general threat. Ned may and often should consider how he came by a hypothesis. But in the end, what matters is not where it came from but whether, and if so why, he is prepared to stand behind it.

Luminosity

What is required for an agent to be in a position to stand behind a commitment? It is plausible that he must know both what the commitment is and that the commitment is his. This might seem obvious. Without awareness of our own mental states, we might think, we would not know or understand anything. But this claim is ambiguous. It might mean that if we were not conscious organisms, we would not know or understand anything. This is likely true, even trivially true. "What is it like to be a turnip?" would probably be neither as long nor as interesting as "What is it like to be a bat?" (see Nagel 1974). Or it might mean that to be epistemically committed to φ, an agent must be consciously aware of that commitment as hers. In that case, epistemic attitudes would be, to use Williamson's term, luminous. If a state is luminous, Williamson says, "whenever it obtains (and one is in a position to wonder whether it obtains) one is in a position to know that it obtains" (2000:13). Williamson focuses on the prospects of luminosity of mental state types—believing, supposing, wondering, and so forth. He argues convincingly that not every token of such a type need be luminous. My concern is with the luminosity of tokens—individual, occurrent states of mind. I focus on cases where not merely is the agent in a position to know that the state obtains, she actually knows or at least reasonably thinks that the state obtains. In my usage, when an agent luminously thinks that the bus is late, she is consciously aware of herself as thinking that it is late. Let us say that when a person's attitude is luminous—when she is aware of herself as harboring it occurrently—she is cognizant of it. Although I agree with Williamson that epistemic states per se are not luminous, I suggest that particular tokens are luminous. And, I will urge, a state's luminosity pays epistemic dividends.

Four positions dominate the current theory of knowledge: reliablism, evidentialism, virtue theoretic epistemology, and knowledge-first epistemology. None holds that knowing that p requires conscious awareness of p or of one's attitude toward p. None, that is, holds that one's epistemic stance toward p must be luminous.

Reliabilism insists that it does not. According to reliabilism, S knows that p just in case S's belief that p is produced or sustained by a reliable process or mechanism (Goldman 1986:48–53). S need not recognize that her belief satisfies this requirement, that the particular process that secures p's epistemic status is reliable, or even that she harbors the belief that p. That knowledge that p is consonant with such obliviousness is touted as an advantage of reliabilism. Because luminosity is not required, reliabilism readily accommodates perceptual knowledge, dispositional knowledge, and knowledge consisting of reliably produced or sustained subliminal beliefs. Its indifference to luminosity also enables reliabilism to recognize that some animals who are incapable of metacognition nevertheless know.

Evidentialism is a bit more coy about the issue. According to evidentialism, S knows that p just in case she has a non-fortuitously justified true belief that p, where satisfaction of the non-fortuitousness requirement ensures that the belief's justification and its truth maker properly align (Adler 2002:164–167). For our purposes, the nature of the alignment is irrelevant. What matters is that evidentialism does not require that S be aware she believes that p, aware that her belief is sufficiently justified, or aware that what justifies it is what makes it true. All that is required is that the conditions in fact be satisfied.

Virtue theoretic epistemology holds that for S to know that p, her belief that p must display her exercise of epistemic virtues (Sosa 1991:225–244). She need not recognize that she believes that p. Nor need she appreciate that her belief is epistemically virtuous. Indeed, to the extent that epistemic humility is one of the relevant virtues, she probably should be ignorant of her belief's manifesting virtue.

Knowledge-first epistemology expressly denies that luminosity is required for knowledge. When one's mental attitude vis-à-vis p is not luminous, then even if S knows that p, her knowledge is opaque to her (Williamson 2000:11–18). Nevertheless, she knows because she stands in an appropriate relation to the facts. Since knowledge is held to be more basic than belief, if she knows that p, questions about belief do not even arise.

It comes as no surprise that knowledge is consonant with such obliviousness. Two familiar factors combine to explain the possibility of subliminal knowledge—knowledge that the agent cannot self-ascribe. The first is the widespread recognition that the KK thesis is false. It is possible to know that p without knowing that one knows that p. Were this not so, then to know that p, S would have to know that she knows that p; to know that she knows that p, she would have to know that she knows that she knows that p, and so forth ad infinitum. We are not that smart.

The second factor is that we have many beliefs without being aware that we have them. This is our ordinary condition with mundane beliefs that do

Epistemic Autonomy

not reach the threshold of conscious awareness. People surely believe that cats don't grow on trees, even though they have never entertained the question. If asked, we would immediately, confidently, and without deliberation be able to voice our opinion. We have a vast array of dispositional beliefs that, until the questions arise, we are unaware of. It is no surprise that they are not luminous. What of occurrent beliefs? Perhaps surprisingly, they need not be luminous either. On "automatic pilot," her mind on other things, Sarah steps around the puddle in the path, not even consciously registering that it is there. All four accounts of knowledge would claim that in such circumstances Sarah knows that there is a puddle in the path and that her knowledge rationalizes her behavior. It is even possible to have beliefs, both occurrent and dispositional, that one sincerely denies having. According to psychotherapists, this occurs in repression; according to Marxists, it occurs in false consciousness. Many such beliefs satisfy the conditions on knowledge. Some patients who harbor repressed memories of childhood trauma subliminally believe and know that they were abused; some workers employed by a seemingly benevolent employer subliminally believe and know that they are exploited. If anything close to contemporary theories of knowledge is correct, knowledge does not require cognizance—that is, explicit awareness—of one's doxastic or epistemic state.

Turn now to understanding. Propositional understanding is expressed in a sentence of the form "S understands that p" or "S understands why p," where the direct object of the sentence expresses a proposition. It is understanding of an individual matter of fact. Ted understands that the house burned down; Tim understands why the bus is late. Objectual understanding is understanding of a topic or subject matter. Susan understands thermodynamics; Paul understands the fall of the Roman Empire; Eric understands the candidate's campaign strategy (see Kvanvig 2003:191–192).

Propositional understanding is a form of knowledge. "S understands that p" seems to be just another way of saying "S knows that p." That being so, understanding that p no more requires awareness than knowing that p does. What of understanding why p? It might seem that understanding why a proposition is true is sufficiently complex that cognizance would be required. Surely, one might think, an epistemic agent can't understand why the house is on fire without being consciously aware of the cause of the fire. But the issue is not so clear. Suppose a firefighter rushes into a house to extinguish a kitchen blaze. He does not turn on his hose. He sprints past an available water source. At considerable danger to himself, he gets close enough to the fire to throw a flame-retardant blanket over the blaze, depriving it of oxygen. Given his expertise, he immediately understands what sort of fire it is. He recognizes that, since it is a grease fire, it would be exacerbated rather than extinguished if doused with water.[3] He understands why that particular fire is burning. No doubt, he is cognizant of the fire. Nevertheless, in extinguishing it, he is on autopilot in much the way the pedestrian who stepped around the puddle was. His understanding, it seems, is embedded in beliefs that do not, and in the circumstances need not, rise to the level of consciousness. People who regularly confront emergencies—first responders, battlefield medics, triage nurses, and the like—regularly display this sort of immediate, subliminal understanding of why an alarming incident is happening. Even if, as Grimm (2014) and others maintain, propositional understanding is knowledge of causes,[4] and even if knowledge of causal connections is more complex than knowledge of (some) brute facts, propositional understanding does not always involve cognizance of the facts one understands or of one's epistemic state as a state of understanding them.

Objectual understanding is holistic. It is a grasp of a systematically linked body of information that is grounded in fact, is duly responsive to evidence, and enables nontrivial inference and action regarding the phenomena the information pertains to (Elgin 2017:44). This certainly seems complicated enough and demanding enough that cognizance would be required to achieve it. I think not. If we focus on cases like understanding thermodynamics or understanding the fall of the Roman Empire, cognizance seems mandatory. It is hard to imagine harboring and exploiting vast networks of scientific or historical information subliminally. But if we consider an adroit basketball player who adjusts his play to accommodate the moves of the opponent he is guarding, the conviction that someone with objectual understanding must have real-time conscious awareness of the factors that constitute his understanding looks suspect. The player, it seems, viscerally understands his opponent's moves. Someone like the firefighter or the basketball player, who has merely subliminal understanding, may be at a loss if asked, "Why are you doing that?" even though he is doing something that is manifestly responsive to the subtleties of the circumstances and effective in dealing with them. Objectual understanding can be embedded in know-how that resides below the threshold of cognizance.

This raises the question: What does luminosity add? Why, if at all, is an agent epistemically better off if she is cognizant of her attitudes? I suggest that to the extent that an agent's attitudes are subliminal, she is at their mercy. Even if she is on objectively solid ground, she is subjectively vulnerable. She is in no position to recognize the strength or structure of her cognitive commitment or network of commitments. Nor is she in a position to evaluate that commitment or network. The firefighter could probably offer a post hoc justification for his action, grounding it in his understanding of the different profiles that different sorts of fires display. He could, that is, answer the question "Why *did* you do that?" by bringing to conscious awareness something he tacitly drew on when acting. But if the action was grounded in a subliminal understanding of the fire, then at the time he acted, he would have found it difficult to answer "Why *are you doing* that?" and not merely because he had more pressing concerns. Moreover, his post hoc explanation may be relatively shallow; it may fail to reflect the intricacies of the understanding that his action displayed. His accommodation to the vicissitudes of the fire may have been as fine-grained as those of the basketball player. Subliminal understanding is real understanding. Nevertheless, cognizance seems mandatory if one is to provide a justification or an assessment. Critical scrutiny requires cognizance of whatever one is scrutinizing. Here I will focus on objectual understanding.

An understanding of a topic consists of a network of epistemic commitments in reflective equilibrium: its elements are reasonable in light of one another, and the system as a whole is as reasonable as any available alternative in light of the epistemic agent's antecedent commitments. Such an equilibrium is labeled "reflective" because together the two features make it a network of commitments that the agent can, on reflection, accept. I will have more to say about this in chapter 4.

When it comes to subliminal networks (or subliminal regions of networks), things are a bit tricky. The problem arises from the word "can." Clearly the agent cannot on reflection accept a consideration if it is inaccessible to her reflection. She cannot scrutinize it, cannot assess it, cannot examine it at all. But if the criteria for reflective equilibrium are satisfied, she understands the topic. We accommodate the worry about subliminal understanding by introducing modal gloss to "on reflection": if she could reflect on it, she would endorse it. By her own lights, there is nothing wrong with her network of commitments per se; what is missing is her capacity to appropriately access that network. This is why subliminal understanding is often effective. The elements of the network are mutually supportive, and the system as a whole is reasonable. That explains why in epistemically resistant cases of false consciousness the network of commitments does not qualify as an understanding. In a resistant case, if the subject could reflect on what she subliminally believes, she would not accept it. She would reject it in favor of the view she consciously holds—namely, that her employer is benevolent. This is a form of confirmation bias.

Understanding is open-ended. A subject's fabric of commitments concerning a topic expands as it incorporates new propositions, new standards, or new methods. It deepens when she integrates new connections among commitments she already holds. A math student deepens her understanding of complex numbers when, having already accepted a proof of the fundamental theorem of algebra, she incorporates a new proof into her network of mathematical commitments. She understands the theorem's role in mathematics better when she sees how the new proof highlights a different network of mathematical relations. Understanding is refined when, on the basis of the commitments the agent already accepts, she comes to draw finer distinctions or relocate boundaries between individuals or kinds. It is strengthened via contraction when problematic commitments are rescinded. Because it is fallible and open-ended, understanding is flexible and dynamic. It alters its constitution and contours over time as it accommodates or rejects new inputs. A network of commitments ought never be construed as fixed and final.

As we have seen, understanding can be subliminal. To the extent that it is, the agent relies on it as a basis for inference or action without appreciating why, or even that, she does so. That, I suggest, leaves her vulnerable. Although she alters her commitments in response to new inputs, adjusting them in the face of whatever impediments she encounters, she does so blindly. For all she can tell, the restored equilibrium—if it actually is an equilibrium—is unstable. Later understanding is not always greater understanding. An adjustment that enables her to escape an immediate pitfall may create more serious difficulties downstream. If the agent cannot survey her network of commitments, she cannot entertain alternatives, spot vulnerabilities, or make corrections.

The epistemic situation of someone whose understanding is subliminal is less than ideal. Unsurveyability engenders epistemic insecurity, even if that insecurity is itself hidden from view. A person whose understanding

Epistemic Autonomy

is subliminal cannot identify the strengths and weaknesses, opportunities and obstacles that figure in her thinking about the topic. She may be on safe ground, but she would be ill-advised to think she is, particularly if the stakes are high. Moreover, she cannot exploit her understanding. Inasmuch as the network is flexible and open-ended, it is responsive to new inputs. But she is in no position to seek out the sorts of inputs that would be fruitful or to safeguard her understanding by preventing the incorporation of inputs that might lead her astray. She can strike out randomly, act on her untethered hunches, or wait for whatever happens to impinge on her cognitive system. She cannot then leverage her current understanding to gain a better understanding.[5]

Cognizance

A luminous understanding of a topic is an explicit awareness of how the elements of a network relate to one another, how the network answers to the evidence that supports it, how it represents its target. A cognizer, consciously aware of what she thinks, can act to fill gaps in her understanding. Recognizing that her mastery of ancient Greek does not enable her to decide whether "episteme" means "knowledge" or "understanding," she can run through her memory of actual uses of the term and see which translation fits best. Failing that, she can run through her list of acquaintances with greater expertise in Greek and ask their opinions. Rather than just waiting for bits of data to conveniently coalesce into an epistemically useful package, she can garner additional evidence. She can marshal antecedently isolated considerations to support a hypothesis. Her evidence may be scattered across her networks of commitments. Jane seeks to understand why Matt concealed his illness. She has been his colleague and casual friend for a number of years. Having attended meetings with him, she knows that he is diffident. Having played on the company softball team with him, she knows that he presents himself as hale and hearty. Having driven his kids to school, she knows that they consider him invincible. So long as these bits of information are unconnected, she finds his reluctance mysterious. But once she puts them together, she sees that collectively they constitute a pattern of strong evidence that he would not want others to feel sorry for him (see Lewis 1983:278–279).

An agential cognizer not only cannot only recognize tensions in her commitments she can also endeavor to alleviate them. On surveying her political commitments, Jill comes to recognize that her commitment to freedom of expression is in tension with her views about hate speech. She believes that people should be free to express their opinions, and also believes that people should not be subject to hateful, derogatory rants criticizing them for their race, gender, sexual orientation, or religion. If she were a purely spectatorial cognizer, she would simply conclude that the tension between her commitments is a regrettable weakness in her understanding of political rights. As an agential cognizer, however, she can and should ask herself, "What should I do about this tension?" She can assess her resources for resolving the dilemma. If she thinks that she currently has no acceptable way to relieve the tension, she can consider where she might go to gain additional resources. Should she read more about the issue? If so, what? Should she consult with others? If so, who? An agential cognizer thus can, in principle, recognize the limits of her current understanding and make moves directed to extending her epistemic range.

The spectatorial stance is third-personal. A spectatorial agent can apprehend the warp and weft of her tapestry of commitments. She can identify her various commitments, recognize their interconnections and lacunae, locate redundancies and idle wheels, as well as putatively relevant commitments that seem not to mesh. She can survey and explore the network. So far, this is purely descriptive. Her epistemic stance is like that of a cartographer, mapping the terrain, identifying and classifying what he finds.

She can do more. Like the sportscaster, she is in a position to explain how and why various commitments hang together. She not only appreciates that the network constitutes a mutually supportive system she can recognize what supports what and why some seemingly relevant items are not included. Perhaps, for example, her survey reveals that, given the standards of evidence the network incorporates, anecdotal evidence or hearsay must be excluded. Then, she can appreciate why what might have seemed like a relevant bit of information ought not be incorporated into her understanding of a topic.

The spectatorial cognizer can assess her network. She can identify gaps and flaws in her understanding. A constellation of mutually supportive commitments may be incomplete. Perhaps her current understanding of the topic enables her to see that there are questions she ought to be able to answer but cannot or that there are informational gaps that she lacks the resources to fill. She may be able to identify errors—commitments that she previously incorporated into her evolving understanding that do not stand up to scrutiny.

Epistemic Autonomy

She can spot weaknesses. Perhaps some supporting links in her network are tenuous. So long as they hold, the network is in reflective equilibrium. But she may appreciate that her reasons for accepting them are relatively fragile. For example, her epistemic support for a commitment might stem entirely from the testimony of a single source or the evidence of a single experiment. That now strikes her as risky. She can discern vulnerabilities. If a proposition she has integrated into her network has a large margin of error, she may see that, despite the antecedent support and despite its contribution to the network, it still a weak element. It does not contribute much. She can recognize obstacles. Perhaps there are limitations in her inferential commitments that prevent her from drawing certain inferences that she would like to be able to draw.

The fruits of her survey are not all bitter. She can also recognize strengths in her network. She can appreciate how and why it constitutes a solid take on the topic. It explains things she wants to explain, provides the resources for inferences she wants to make, draws lines between cases where it seems important to draw them. She can see how her current understanding supplies resources to extend her network of commitments and to engage in effective actions.

If the spectatorial cognizer had access to it, she could survey someone else's understanding of the subject in exactly the same way. She could identify the strengths and weaknesses, opportunities and obstacles to his understanding of the topic. He ought to be able to explain, infer, extrapolate, exploit, and so forth. He is in a position to avoid these pitfalls, draw these inferences, extrapolate from those findings. He ought to shore up this commitment and revise that one. Inasmuch as the spectatorial stance is third personal, this is not surprising.

Someone adopting a spectatorial stance can recognize where and how things might be better; he can provide cogent reasons for the verdicts he reaches. But having taken a third-personal, outsider stance, he cannot implement the needed changes. An anecdote of John Perry's brings this out. He says:

> I once followed a trail of sugar on a supermarket floor, pushing my cart down the aisle on one side and back the aisle on the other, seeking the shopper with the torn sack, to tell him he was making a mess. With each trip around the counter the trail became thicker. But I seemed unable to catch up. Finally it dawned on me. I was the shopper I was trying to catch. I believed at the outset that the shopper with the

> torn sack was making a mess. And I was right. But I didn't believe that I was making a mess. . . . When I came to believe that, I stopped following the trail around the counter, and rearranged the torn sack in my cart. (1979:3)

Initially, Perry adopted a largely spectatorial stance toward the mess making. He thought that the trail of sugar was likely caused by a shopper with a leaky sack in his cart. He inferred that the shopper would probably not want to be spilling sugar and that the shopper would therefore want to know that he was making a mess. Perry did not follow the trail merely because he wanted to know where it led but rather because he sought to intervene to prevent further mess making. To that limited extent, his stance was agential. But because he did not know that he himself was making the mess, the best he could hope to do was to catch up with the mess maker and recommend that he adjust the leaky sack. It was only when Perry realized that he himself was making the mess that he could reposition the torn sack so that it no longer spilled sugar onto the floor. That is, once he recognized that he himself was the mess maker and adopted an agential stance, he could resolve the problem.

Agential understanding is fundamentally first personal. Someone whose stance is agential can take responsibility for her networks of cognitive commitments. Like the spectatorial cognizer, she can survey and assess her understanding of a topic. She can identify its strengths and weaknesses, recognize the opportunities and obstacles to further understanding that it presents, identify errors and vulnerabilities embedded in it. But unlike the purely spectatorial cognizer, she can do something about what she finds. She can amend, rescind, or strengthen those of her commitments that she now considers under-supported or ill-advised. She can also intervene in ways that a mere spectator cannot. There is, of course, no guarantee that she will thereby improve her understanding. That depends on the circumstances she is in and the changes she chooses to make. But she is capable of making intentional changes in response to perceived inadequacies and of taking advantage of what she sees as opportunities. If, for example, she recognizes that a consideration or technique that strengthens support in one area could be effectively used to shore up another, she can import it. An agential cognizer does not merely exploit opportunities when they appear. She frames her conceptions of the phenomena in ways that admit of intervention.

Thus far, I have discussed the ways cognizance enables an agent to exploit her views about her target. This is not the whole story, for cognizance is Janus faced. To be able to exploit the resources her understanding provides, an agent

Epistemic Autonomy

needs to be self-aware. She needs to recognize that she herself is capable of taking steps to improve her epistemic lot. This involves recognizing that she herself is in circumstances where certain actions or inferences are called for and that she herself is in a position to undertake these actions or draw these inferences. The "she herself" locutions are designed to underscore that her stance toward her situation must be first personal. It cannot be the disinterested perspective of thinking merely that someone should do something about the problem. The epistemic agent qua agent requires Leibnizian apperception—"reflective knowledge of [an] internal state" (Leibniz:1989§4:208).

Think back to Perry's predicament. He was cognizant of the fact that someone was making a mess and consciously supposed that the perpetrator probably did not want to make a mess. He realized that he suspected that the perpetrator, if alerted, would take pains to cease his mess making. That is, Perry was aware of his own views about the spilled sugar and the sugar spiller. He also recognized that he himself was aware of mess making in his vicinity—aware that, for example, he was not looking through a surveillance camera at mess making in a supermarket across town. He recognized that he himself was in a position to alert the perpetrator. He was not, for example, paralyzed or incapable of speech. And he thought that a good way to find the perpetrator was for him himself to follow the sugar trail. In order to undertake the task, he had to be cognizant of his own situation, with its affordances, powers, and limitations, not just the perpetrator's.

Luminosity is epistemically valuable because it shows an agent where she stands. It provides the resources that enable her to identify and assess the obstacles and affordances she faces. She can appreciate how solid or shaky her epistemic ground is. This enables her to exercise her autonomy—to use, extend, or refrain from using considerations as she sees fit. Luminosity reveals what resources the agent has to build on and, by her own lights, what risks she takes in building on them.

To be in a position to exploit his resources, the cognizant agent must also be able to take a normative stance toward his network and his resources for improving it. I said earlier that an agential cognizer can do more than assess his commitments as a sportscaster would; he can also take steps that he has reason to think will improve his epistemic situation. Both a spectator and an agent have the resources to recognize that their network includes an epistemic infelicity—perhaps an invalid inference license that permits affirming the consequence. Both can recognize that this is a defect. But the agential cognizer is in a position to devise and test remedies. To do that, he needs the capacity to be not only critical but also self-critical. He needs to be able to take a normative stance not just to his understanding but also to himself as someone who understands. He is then in a position to reflectively endorse those of his commitments which stand up to his scrutiny.

The requirement of reflective endorsement does not leave the agent free to decide however he pleases. His revisions should, as far as he can tell, strengthen his network of commitments. Minimally, he should constrain himself to accept only lower-order considerations that, as far as he can tell, are vindicated by higher-order principles that he accepts, and accept only higher-order principles that mesh with and provide a rationale for the lower-order considerations he accepts, while providing grounds for repudiating the lower-order considerations he rejects (see Goodman 1983:63–65). Adjudication may be required to bring his various commitments into accord. Strong consistency and coherence conditions constrain what he should accept. They require that, as far as he can tell, like cases are treated alike, that the considerations he accepts be jointly, not just individually, acceptable, and that the methods and standards he uses actually vindicate the conclusions he draws from them. The constraints block such epistemic indiscretions as making snap judgments, appealing to false equivalences, jumping to conclusions, and special pleading. This is all to the good. Nevertheless, the process may still seem woefully subjective. In accepting considerations that he reflectively endorses, the agent satisfies his own standards. But he still seems vulnerable to systematic errors that stem from lax or skewed standards, such as confirmation bias, tunnel vision, neglect of base rates, and so forth. If epistemic autonomy leaves him in such a sorry state, it is not much of an asset.

Kant Redux

To escape this predicament, again it pays to turn to Kant (1981). He characterizes an autonomous agent as someone who acts on laws he makes for himself. But from the fact that he makes the laws *for* himself, it does not follow that he makes them *by* himself. He does not. Although Kant takes the categorical imperative to apply exclusively to ethics, by extrapolation, we can secure the basis for epistemic normativity (Elgin 2017:105–111). Kant provides multiple versions of the categorical imperative, three of which concern us here. Under extrapolation, each reveals something important about

Epistemic Autonomy   51

epistemic acceptability. My extrapolation makes no transcendental assumptions, nor does it deliver any transcendental results. It is purely pragmatic. By tracing the trajectory through the various versions of the categorical imperative, I show how to move from the subjective to the intersubjective in a way that exploits the autonomous agent's strengths while minimizing his vulnerabilities.[6]

A "maxim" for Kant is a principle on the basis of which an agent acts (1981:§400). Since accepting is an instance of acting, the principle according to which an epistemic agent accepts qualifies as a maxim. The initial formulation of the categorical imperative (CI_1) is the principle of universalizability: act only on a maxim that you could simultaneously will to be a universal law (1981:§421). It precludes making an exception for your own case. By extrapolation, an epistemic agent should accept a consideration only if it would be universally acceptable. For now, let us bracket the question of how wide the relevant universe needs to be. I will return to it below. Whatever the answer, if the evidence is good enough to provide sufficient reason for Ned to accept p, it is good enough to provide sufficient reason for his compatriots to accept p as well. The responsible epistemic agent thus recognizes that the principles on the basis of which he accepts or rejects a consideration are ones that other members of his community ought to countenance too.

It might seem that this brings us no further. If an agent reflectively endorses p, he accepts p and thinks that his reasons for accepting p are sufficient. That being so, he naturally thinks that other epistemic agents should accept p. After all, he thinks, he is right! Because p is correct, his compatriots ought to accept it. No real appeal to them is necessary. Universalizability apparently comes for free. CI_1 may seem to invite this conclusion.

The second formulation of the categorical imperative (CI_2), the principle of humanity, blocks it. "Act in such a way that you treat humanity, whether in your own person or in the person of another, always as an end, never merely as a means" (Kant 1981:§429). Here, the focus is on agents rather than maxims. To treat others as ends in themselves is to treat their perspectives as worthy of respect—as worthy of respect as one's own. So the agent cannot simply conclude that others ought to accept p because he has established to his own satisfaction that it is correct. Rather, to satisfy the epistemic analog of CI_2, he should ascertain whether, from their perspectives, p appears acceptable. Only if it does ought he accept it. Still, there is a worry. CI_2 seems to say nothing about the basis of unanimity. Should he accept a contention just because his compatriots happen to think the same, regardless of why? The agreement might just be evidence that the contention is popular.

The third formulation (CI₃), integrating the previous two, answers that question. An agent should act only on a maxim that she can endorse as a legislating member of a realm of ends (Kant 1981:§431). This requires explication.

A realm, Kant maintains, is "a systematic union of different rational beings through common laws" (1981:§433). Because members of a realm of ends make the laws they are subject to and consider themselves subject to the laws precisely because they have made them, the satisfaction of the categorical imperative is a manifestation of autonomy.[7] The acceptability of a consideration is dependent on the capacity of each member to reflectively endorse it. Let us call the members of a realm of ends "compatriots."

Kant characterizes the agent as a legislator, not an autocrat. Enacting a law is different from issuing an edict. There are procedures that must be followed—procedures that the legislators collectively contrive and reflectively endorse because they believe that those procedures will further their legislative purposes. Not being an autocrat, an agent cannot secure his conclusion simply by decreeing, "Because this is what I think, everyone else should think it too! That settles it!"

A realm of ends has multiple members. The agent is *a* legislator, not *the* legislator. An absolute monarch can issue edicts on his own. Legislators enact laws collectively. So if he has any hope of succeeding, the consideration that an individual legislator advances should be acceptable from the points of view of the other legislators. That is what universalizability requires.

To enact a law that he favors, a legislator has to convince other lawmakers that the proposed legislation is acceptable *from their points of view*. A point of view or perspective is an epistemic agent's subjective take on things. It consists of the network of commitments he is personally, on reflection, inclined to accept. The legislator thus has to convince his compatriots by appealing to considerations that are acceptable from their points of view. He cannot permissibly bully or bribe or befuddle them into submission. It is not enough that he thinks it is a good idea. Nor would it be enough if each of them independently happened to think it a good idea. The fact that they agree with one another and the reasons why they agree with one another are factors in their reflective endorsement. Each should accept it at least in part because it is acceptable from the points of view of the others. Each should treat the others not just as sources of evidence but as co-assessors, who deem the

Epistemic Autonomy

evidence acceptable from their own perspectives.[8] Ned thus has to engage with his co-legislators and appreciate how things look to them. To make his case, he needs to articulate reasons that they can understand and accept. Only if a consideration the agent advances stands up to scrutiny from other perspectives will it pass muster. Individual legislators recognize that their perspectives are limited and may be skewed. The deliverances of their several perspectives may initially be at odds. Their goal is to adjudicate, not merely to accommodate whatever anyone happens to think. It follows that only if a consideration stands up to scrutiny from other perspectives should an agent hold that it stands up to scrutiny from his own. If he is wrong in believing that it is acceptable from their points of view, he is wrong about what he ought to accept. A factor in its being acceptable to an epistemic agent from his point of view is that it is acceptable to his compatriots from theirs.

So far, I have at best shown how we might construct a conception of epistemic acceptability by extrapolating from Kant. But why should we accept it? By stipulation, the agent's relevant ends are cognitive. He wants to know, understand, or grasp how things genuinely are, not just how they seem to him. He is cognizant of some of his limitations—some of the ways his evidence, methods, or standards of assessment might systematically lead him astray. And he appreciates that if they systematically lead him astray, he may, on his own, lack the resources to identify or rectify them. He recognizes that he is not alone in being so limited. Many of his limitations are apt to afflict any individual epistemic agent. So it is reasonable for him to attempt to shore up his commitments—even those that seem unobjectionable from his point of view. And it is reasonable for him to do so in a way that simultaneously shores up the commitments of his compatriots. Counterparts to the several formulations of the categorical imperative afford resources for doing so. A universalizability principle akin to CI_1 holds that whatever considerations justify him in accepting p equally justify others in accepting p; whatever considerations justify him in rejecting q equally justify others in rejecting q; and whatever considerations justify him in withholding r equally justify others in withholding r. An analog to CI_2 maintains that it is not enough that the agent, from his perspective, think that p is justified. Others, from their own perspectives, should think so too. That is, it is not enough that were others to adopt his perspective, they too would think that p. They must be able to accept it from diverse perspectives. And an analog of CI_3 insists that the grounds for their agreement should consist of considerations that they could also reflectively endorse on the basis of principles that they individually and collectively consider to set reasonable constraints on epistemic acceptability. This yields an *epistemic imperative*. "An epistemic agent should accept only considerations that he could reflectively endorse as a legislating member of a realm of epistemic ends" (Elgin 2017:105).

Raising Objections

I said earlier that epistemic autonomy and interdependence are mutually supportive. It might seem that I've mainly argued that the autonomous agent gains from the support of an epistemic community. Not only does he gain access to additional information via testimony and additional skills via instruction, community support also stabilizes his commitments by giving him reason to think that they can be sustained from diverse points of view. But the benefits must go both ways. A community supports epistemic autonomy by underwriting principles that are tenable only if generated, ratified, and endorsed by autonomous agents—agents who would and could withhold acceptance if, from their perspectives, the principles seemed untenable. This will only work if the individual actors are ready, willing, and able to withhold their acceptance.

Unless an actor is autonomous, he has nothing to contribute. Perhaps, desiring their approval, the heteronomous actor just reiterates the commitments of his compatriots. He goes along to get along. In that case he is an epistemic parasite. If other members of the community cannot be confident that he would raise objections if a consideration was not supported from his perspective, his agreement counts for nothing. Because he is just a yes-man, the community gains nothing from his affirmation. He would have agreed with anything the powers that be maintained. Perhaps, on the other hand, he goes his own way, ignoring his compatriots' views completely. Again he contributes nothing. Either way, the community's epistemic position is weakened. A perspective that might in principle have something to contribute is left without an advocate. Actors must be epistemically responsible to constitute a realm of epistemic ends.

No doubt, an epistemic community can get by with a few free riders. But unless most of its members are autonomous and count on one another to act autonomously, consensus counts for little. Indeed, interdependence is disastrous if the community is cognitively corrupt. If they merely parrot one another's convictions or the convictions of their leader, their yea-saying does not supply independent support. Unless the individual autonomous agent draws on the support of a realm of epistemic ends, he is epistemically vulnerable. He may be subjectively secure in that he satisfies standards he sets for himself. But he has no way to ensure that he is not prey to weaknesses that his parochial perspective cannot disclose. Unless the epistemic community is composed of autonomous epistemic agents, consensus affords no basis for confidence. The grounds for agreement may be spurious. Epistemic autonomy and interdependence are thus mutually supportive. Each is too weak to sustain acceptability without the other.

Epistemic Communities

Although Kant takes the moral realm to include every rational agent throughout history, his characterization allows for realms that are more restricted. This is mandatory if the extrapolation to epistemology is to be remotely plausible. Few if any epistemic commitments would be acceptable to all rational agents, regardless of time, place, experience, or education. Nevertheless, more restricted communities qualify as systematic unions of different rational beings through common laws (Kant 1981:§433). We speak of, for example, the community of plasma physicists or the community of economic historians. It is plausible to construe the members of a discipline as composing a self-constituted realm bound together by cognitive commitments that they take to govern and circumscribe their professional lives. These are their common laws. There are more informal realms as well—for example, the guys down at the pub who deem one another's opinions about football worth listening to, even when those opinions diverge. They tacitly accept standards for what sorts of considerations deserve to be taken seriously. A realm is epistemic when the considerations its members consider worthy of collective endorsement are cognitive. They are grounded in the goal of grasping how things stand.

The suggestion that we are legislating members of such communities may seem unrealistic. We regularly find ourselves answerable to demands that we had no role in contriving and see no hope of changing. No doubt, we often feel powerless in the face of the demands of an entrenched epistemic community. Nevertheless, at least as an idealization, this picture might seem plausible for adults. It may take time and effort, but epistemic agents can successfully mount challenges and eventually get the standards, norms, and practices of their epistemic communities changed. The emergence of requirements that experiments have controls, that journal submissions be blind refereed, and that auditions be held behind a curtain were products of sustained efforts to improve epistemic standards in order to ensure that irrelevant factors do not influence decisions. Similarly, with time and effort, we can bring it about that the community recognizes the merits of hypotheses it was initially inclined to reject. Think of the emergence of a consensus that peptic ulcers are caused by a bacterium rather than by stress. Although there may be considerable resistance to challenge, that resistance can be overcome.

The difficult case here concerns the young student. However much we idealize epistemic communities, there is virtually no chance that a ten-year-old history student can buck the standards that will be used to grade his exam. Rather than a legislating member of a realm of epistemic ends, it may look like he is simply a slave to the demands of an epistemic autocracy.[9] We might want to justify his powerlessness by appealing to the idea that children are not yet qualified to function as legislating members of realms of ends. They are not yet fully autonomous. We could make an analogous point about laymen who lack the expertise to buck the standards of communities of inquiry they do not belong to. Non-engineers are ill equipped to contribute to debates about the acceptability of methods for measuring metal fatigue and ill equipped to challenge the metallurgical standards they rest on. They do not qualify for membership in the relevant realm of epistemic ends. This argument is plausible and often correct. But the child ought not be construed as a mere slave to an epistemic autocracy. He is a legislating member of a different realm of ends—one consisting of his fellow students. Being conscientious, he appreciates that by accepting a consideration, he gains the right to convey it to his peers on terms they can reasonably endorse. He accepts the claim that the British won the Battle of Trafalgar on the grounds that that's what his history book says, and he rightly thinks that the history book's say-so satisfies his classmates' standards for the sort of information they need to prepare for the test. Thus, he can responsibly report to his peers that the British won because he recognizes that that information satisfies a relevant standard that they share. The standard, in this case, need be no more elevated than like "good enough to get us through the test." But he appreciates that the popular opinion among his peers that the victory's being portrayed in a video game is not good enough.

Epistemic Autonomy

Epistemic autonomy and epistemic interdependence go hand in hand. Autonomy without interdependence leaves an agent in a precarious position; she has no assurance that her convictions are free from biases that are invisible from her point of view. Interdependence without autonomy leaves the community in an equally precarious position. It has no assurance that consensus is not a product of epistemically irrelevant or even epistemically pernicious features. But a community comprised of autonomous agents, functioning as such, provides grounds for epistemic acceptability. There is no assurance that their conclusions are true, but there is assurance that they are reasonable in the epistemic circumstances.[10]

3 Epistemic Engagement

Epistemic Responsibility

I have argued that members of a realm of epistemic ends are ineluctably interdependent. Each depends on information she gleans from others, methods she learned from others, instruments that were designed and calibrated by others. Each contributes to the common store. A critical question is what makes that information and those methods trustworthy. It is not enough that they are in fact reliable. Members of the community need reason to credit them. That requires that the members of the community both be and expect their informants to be intellectually responsible. Members of a realm of epistemic ends must be trustworthy in their epistemic dealings with one another and must be justifiably confident that their compatriots are also trustworthy. A presumption that the agent is epistemically responsible is the ground for trust. It is not an add-on. It is built into the nature of information transfer. That being so, it deserves explication.

Informative exchanges are cooperative (Grice 1989). Parties attune their contributions to the (often-evolving) goals of that exchange. This gives a conversation a focus and a direction, distinguishing it from a random collection of remarks. Informative exchanges are governed by maxims—rules endowing joint expectations with normative force. Because participants mutually recognize that they are bound by the maxims, they interpret another's contributions as either satisfying the maxims or flouting them in a way that manifestly fosters the goal of the exchange. We might, of course, dispute any or all of Grice's specific maxims.[1] What is important is that informative exchanges rely on jointly recognized, normatively binding rules that, like rules of a game, circumscribe what moves are permissible as the exchange proceeds. They set constraints on precision, relevance, evidence, even tone.

Although Grice characterizes his maxims as constraints on conversation, this is slightly misleading. First, in conversations such as flirting and conveying condolences, they do not apply. Despite the maxim of quality, which requires that contributions be informative, saying something obvious, hence uninformative, in such cases can be entirely appropriate. It wouldn't do to hold that an utterance of "I love you" is appropriate only when the issue is in doubt. Moreover, when it is manifest that someone is about to utter something unforgivable, flouting the maxim of relevance by radically changing the subject may be a heroic intervention. Rather than furthering the ends of the exchange, the best bet is to bring that particular exchange to an abrupt halt. Not all conversational exchanges aim or purport to aim at information transfer. When they do not, different norms apply. Second, the maxims apply to modes of communication such as lectures, journal articles, and news reports that are not conversational. The maxims pertain to information transfer, not primarily to the communicative act by which the transfer is accomplished. I suggest that the reason Grice's maxims apply to communication for the purposes of information transfer is that they are in large measure epistemic. They specify mutual expectations that underwrite transfers of epistemic entitlement. Lying and misleading are effective only because compliance with such maxims is the default expectation. Where the goals of participants to an exchange involve the transfer of information recipients can reasonably rely on, Grice's maxims or their close kin apply.

It might seem that the expectation underlying epistemically responsible information transfer is straightforward: an informant should restrict her contributions to statements she considers true. Things are not so simple. People consider all sorts of things true. Superstitions and conspiracy theories abound. The mere fact that someone considers p true is at best a weak reason to think that she ought to contribute it to a communicative exchange that aims to foster the epistemic ends of the participants. The supermaxim of quality, "Try to make your contribution one that is true," subsumes two submaxims: (1) "Do not say what you believe to be false" and (2) "Do not say that for which you lack adequate evidence" (Grice 1989:26). The second maxim is the epistemic lynchpin. Because information transfer conveys epistemic entitlement, the information transferred does not arrive empty-handed. It comes with an assurance that it is backed by evidence.

Let us call the second submaxim the *evidence maxim*. Since parties to an informative exchange accept it, when someone says or otherwise imparts p,

Epistemic Engagement

she indicates to her audience that she has what they mutually agree is adequate evidence for p. If they take her word, they not only accept p they also accept her assurance that she has adequate evidence for it. They acquire epistemic entitlement on the basis of her say-so. Because an agent's relation to reality would be narrow, precarious, and gerrymandered without the support of others, as Linda Zagzebski urges, stable and robust cognitive contact with reality requires that an individual epistemic agent be linked both to the phenomena her views bear on and to other members of her community (1996:167). Given the ways we learn from and depend on one another, respect for the evidence maxim is critical to epistemic interdependence.

Epistemic responsibility involves competence, conscientiousness, open-mindedness, and sincerity. Insofar as these are epistemic traits, they are dispositions to think and act in specific ways. An agent does not qualify as competent, conscientious, open-minded, or sincere on the basis of a single good decision. Nor does she qualify as incompetent, blinkered, cavalier, or insincere on the basis of a single bad one. What matters is how she is disposed to behave.[2] At a first pass, competence is a matter of what one accepts; conscientiousness, a matter of how and why one accepts it; open-mindedness, a matter of receptivity to other points of view; and sincerity, a matter of how (or whether) one registers and/or imparts information. Taken together, they connect the agent both to the phenomena his views concern and to the epistemic community whose standards determine the acceptability of his views. All are complex, multifaceted, and context sensitive. All can be domain specific. An agent could, for example, be competent, conscientious, open-minded, and/or sincere with regard to one topic but not another. He might, for example, be scrupulously honest about the content of his research but routinely exaggerate his romantic successes. In what follows, I explicate the several traits. My goal is to show how they collectively underlie epistemically responsible information transfer.

Competence

To be competent is to possess a skill—a disposition to perform successfully under certain circumstances (see Sosa 2007:29). This says very little until we have identified what it is to perform successfully and what those circumstances are. According to Sosa, competence is indexed to normal circumstances, where "normal" seems to be equivalent to "usual" or "typical." I want to resist this, as in some cases whether a person is competent depends on how he is disposed to perform in highly abnormal circumstances. The difference between a competent and an incompetent battlefield surgeon or test pilot, for example, might depend not on how well each performs in normal circumstances but rather on how well each performs when things go horribly awry. I also diverge from Sosa on the criterion for success. Sosa's model is an archer who qualifies as competent because in normal archery conditions she regularly hits the target. Her performance is apt, he says, when and because the accuracy of her shot is due to her skill. This is a plausible characterization of a successful archer. Owing to her competence, she usually hits the target. But a competent professional baseball player's batting average is about 0.250. Performing successfully should not in general be equated with either literally or metaphorically typically hitting what one aims at.

What does epistemic competence consist in? Since knowledge requires truth, Sosa's characterization of a competent belief as true because apt seems plausible for knowledge. But we go too quickly if we assume that epistemic competence always involves believing the truth. At the frontier of inquiry (and even well back from the frontier), a competent epistemic agent may frequently fail to believe the truth. Indeed, although she accepts considerations as appropriate bases for inference and action, she may, and may properly, fail to form beliefs about the target. Moreover, understanding is consonant with divergence from truth. Scientists understand their targets via models that they know to be false. In what follows, I restrict the use of "epistemic competence" (with "epistemic" sometimes tacit) to the sort of competence that figures in or underlies understanding. It does not always require truth. Indeed, it sometimes is enhanced by suitable divergences from truth.

Epistemic competence consists in having and being able to draw on a tenable network of cognitive commitments about a topic—perhaps fourteenth-century French lute music, the life cycle of the periodical cicada, or the best way to navigate rush hour traffic in Shanghai. Criteria of adequacy are variable, and are keyed to reasonable expectations. A competent sixth grader's understanding of the life cycle of the periodical cicada would not and should not be expected to pass muster if assessed against the standards of a community of professional entomologists. A competent commuter's understanding of Shanghai traffic patterns would not and should not be expected to pass muster with the community of Shanghai ambulance drivers. As we have seen, understanding involves a capacity to assess relevant evidence, to draw

Epistemic Engagement

valid inferences from that evidence, sometimes to act on it, and perhaps to identify and rely on those with relevant expertise. The threshold of adequacy is determined by the legitimate expectations of the relevant community of inquiry (see Goldberg 2018). It is fixed by the level of evidence that other members of the community expect an informant to have in order to responsibly impart information .

Should we say then that competence requires knowledge or reliable belief or (approximate) truth of one's beliefs about a topic? Katherine Hawley suggests something like this, arguing that trustworthiness requires competence, which she takes to be an ability to speak the truth (2019:63).[3] There are several difficulties with such a suggestion. The first pertains to novices—children and neophytes. They can display some level of competence, even if their beliefs diverge considerably from the truth. The young child who thinks that green plants feed on sunlight is wrong. Still, she has a nascent understanding of photosynthesis. To defend a knowledge-based view by arguing that her view is at least approximately true, we would need to be quite generous about where the boundaries on acceptable approximation should be drawn. Perhaps we should be that generous, but doing so would involve a considerable sacrifice of precision and accuracy. A second worry comes from fields where the evidence seems unavoidably sparse. Paleoanthropology has very little evidence about the day-to-day lives of Neanderthals and is unlikely to find much more. So any understanding of Neanderthals seems fated to fall far short of the standards for knowledge or (non-fortuitously) reliable belief. Nor is it clear what would count as approximate truths about such matters as whether Neanderthals believed in an afterlife. A third worry involves the vanguard of inquiry. Whatever is required for competence in, say, string theory or the physics of dark matter, it is not knowledge, non-fortuitously reliable belief, or approximate truth. Scientists are far from achieving any of those milestones. Nevertheless, it is widely held that some scientists in such fields are more competent than others, and that experts are more competent than laymen. The final worry concerns philosophy itself. Some philosophers seem competent. Many of them develop and endorse theories that we consider far from the truth. Some reliabilists consider some internalists competent, and conversely; some consequentialists consider some deontologists competent, and conversely. Moreover, as Fumerton points out, in philosophy there is nothing close to a consensus about any significant issue (2010:109). I will say more about this in chapter 8. If we want to ground assessments of competence in knowledge, reliable belief, or approximate truth, we probably should cease to consider ourselves or even our most gifted philosophical compatriots competent. Indeed, if we accepted such a standard, we would probably have to conclude that there is no such thing as competence in philosophy.

Evidently, a competent epistemic agent must be capable of identifying and appropriately reasoning on the basis of considerations that foster her epistemic ends. She needs a suitably sophisticated understanding of her subject matter. She needs to satisfy Grice's maxims of evidence and relevance, where these are indexed to the legitimate expectations of her epistemic community. Whether or to what extent that requires knowledge, reliable belief, or indeed any sort of belief varies with subject matter, the level of understanding that is called for, and the current level of understanding in her field.

In any case, competence requires more than mastery of a subject matter. A competent epistemic agent needs to be (at least subliminally) aware that she has the requisite mastery. She needs the ability to exploit it—to use it to serve her epistemic ends. If she cannot, her mastery is inert. Consider the predicament of an epistemically insecure student. Because she underestimates her own proficiency, she goes into the exam convinced she will do poorly. Her situation is not just psychologically unfortunate. It may be academically costly as well. Her epistemic insecurity could easily lead her to underperform. Because her lack of confidence leads her to check and recheck every answer obsessively, she might fail to finish the test. Because she does not trust her command of the material, she might be reluctant to make or articulate subtle inferences, to adduce relevant but unfamiliar information, to venture and defend original and powerful hypotheses. Had she been suitably self-aware, she could have breezed through the test. Competence thus requires a level of relevant, justified self-confidence—a (perhaps implicit) recognition that one reaches a suitable threshold for understanding a topic. The competent student recognizes that she has what it takes to do well on the exam. The competent scientist recognizes that his grasp of the subject satisfies the standards of his scientific community. To accept a consideration is to be willing and able to use it for inference and/or action when one's ends are cognitive. One would not and should not be willing to use a consideration, no matter how well founded, unless one recognized, or at least had good reason to think, that it satisfied relevant epistemic standards. And ceteris paribus, one should not be reluctant to use a consideration once one recognizes that it does satisfy those standards.

Epistemic Engagement

The standards, as we have seen, are not merely subjective. Competence requires that the agent be attuned to community standards of evidence, inference, and relevance, and that she be aware of how her grasp of the material fares with respect to them. Such standards fix what members of a given community legitimately expect of one another (Goldberg 2018:145–185). They thus determine the level of support other members of a community justifiably assume a speaker has when she intentionally conveys information to them. Thinking of herself as a member of a given community, whether it be the sixth-grade science class or the community of professional entomologists, an agent should take herself to be answerable to the relevant community standards. Ordinarily, this involves recognizing that the considerations she accepts reach or surpass the threshold for figuring fruitfully in the community's deliberations. Occasionally, it involves a meta-level assessment, which entitles and equips her to challenge those standards responsibly. Either way, competence requires that the agent be cognizant of the standards and of where a consideration stands with respect to them.

The competence required of a responsible epistemic agent thus has three dimensions: understanding of the topic, self-awareness of her grasp of the topic, and awareness of and responsiveness to the standards and assumptions of the relevant realm of epistemic ends.

Conscientiousness

A conscientious epistemic agent is duly responsive to reasons. She gathers and assesses evidence judiciously. In imparting information, she is aware of and sensitive to community expectations and standards. This, as we saw, involves competence. The agent should recognize the sorts of reasons or evidence required for the acceptability of a consideration of a given kind. She should be aware of what evidence she needs if her acceptance and communication is to be responsible. Let us stipulate that the agent is competent in order to investigate the character traits she needs in order to exploit her information.

In my discussion, I use a variety of normatively loaded terms, such as "suitable," "appropriate," and "duly responsive," whose criteria of adequacy I leave unspecified. Let us simply grant that whatever the correct criteria, a conscientious epistemic agent either satisfies or challenges them, where to challenge a criterion is to call it into question on the basis of publicly available reasons that the community ought not dismiss out of hand (see Scanlon 1998). Here too norms are indexed to appropriate expectations. To be duly responsive to reasons, a conscientious botanist must accommodate or challenge the latest available evidence bearing on her hypothesis, and she should use or challenge up-to-date methods in her assessment of that evidence. A fifth-grade student might qualify as duly responsive if he accepts a claim because he read it in his science text, while rejecting a contrary claim because his only source was a work of science fiction. Being ten years old, he can be expected to know that fictional accounts are untrustworthy sources of factual information, but he cannot be expected to check whether the textbook, which his teacher presents as dependable, is actually trustworthy.

A conscientious thinker is attentive to factors that bear on acceptability; a careless thinker is not. One mode of carelessness is capriciousness. A capricious agent jumps to conclusions, with little interest in, and little attempt to assess, the reasons that bear on them. If something seems right, that's good enough for her. She refuses to be weighed down by the burdens of deliberation. Far from being a conscientious epistemic agent, she is an intellectual butterfly. Without being capricious, a careless agent might nevertheless be disposed to negligently overlook relevant factors, slight their significance, adopt a distorting perspective on the phenomena in question, or draw plainly invalid inferences. In a given case, there may be legitimate disagreement about whether factors overlooked are relevant and/or significant, whether the perspective distorts, whether inferences are plainly invalid. What is needed for the contrast here is simply to recognize that sometimes such negative assessments are sound. Sometimes the fault lies in the agent's negligence, distortion, or cavalier disregard for relevant factors. Then the charge of carelessness can be upheld. When it is, the agent is prepared to accept and impart information that does not stand up to scrutiny. In voicing her opinion, she, like the intellectual butterfly, violates the evidence maxim. She imparts information for which she lacks adequate evidence.

Not all failures of conscientiousness stem from indifference to reasons. An epistemically fickle agent vacillates in her commitments, endorsing any and every passing cognitive fad. One day she is a dualist; the next day, a materialist; then, a functionalist. One day she is a theist; the next day, an atheist; then, a pantheist. Far from being indifferent to reasons, she is swayed by pretty much every reason that crosses her path. But her responsiveness to reasons is shallow. She is indifferent to their force or is incapable of assessing it.

Epistemic Engagement

She floats with every prevailing wind. A conscientious epistemic agent needs to be more than merely responsive to reasons—she must be appropriately sensitive and responsive to their force.

Overlooking information is not always negligent. If data are irrelevant, they ought to be ignored. It would be a mistake to rely on them, and potentially misleading to adduce them. It may be a mistake even to register them. Even when data are relevant and important, an epistemic agent's conscientiousness is not impugned if they are inaccessible. Conscientiousness is no epistemic panacea. But epistemically culpable negligence can typically be avoided. A conscientious epistemic agent is normally aware of what the relevant standards are and aware of whether she has satisfied them. That being so, why are epistemic agents culpably negligent?

One reason is that an agent is indifferent to a particular issue. She simply doesn't care whether her opinion satisfies relevant epistemic standards for acceptability. In that case, she may have no incentive to gather and assess evidence judiciously. Having no opinion on a topic and no incentive to form an opinion need not be epistemically blameworthy. If the topic is one the agent has no interest in, then so long as she is not remiss for having no interest in it, remaining indifferent may be an efficient use of her time and energy. To harbor no views about, say, fashion or sports seems innocuous. So to fail to gather and assess reasons that bear on the stylishness of turtlenecks or the frequency of successful onside kicks is unobjectionable. Nor need the issues on which a person is non-culpably indifferent be trivial in some greater scheme of things. To refrain from forming views about the heat death of the universe, on the grounds that one has no interest in cosmology, is not irresponsible. The issue is important, but it is not one on which every responsible epistemic agent need take a stand. The distribution of cognitive labor regularly lets most of us off the hook. Although it is epistemically objectionable to accept a consideration without giving due weight to relevant reasons, except in areas like morality, where indifference is itself apt to be culpable, refraining from forming opinions on topics that are irrelevant to our lives is permissible.

Still, we might fault someone who forms opinions on a topic without regard to reason. To have no opinion about the stylishness of turtlenecks or about the gravitational effect of dark matter is unobjectionable. But to have opinions on such things without acknowledging the force of reasons that bear on those opinions seems epistemically irresponsible. It would surely be irresponsible to venture such opinions so as to implicate that one had the backing of appropriate reasons. To do so would be to impart information that, so far as one can tell, does not stand up to scrutiny. That would violate the evidence maxim.

At the opposite extreme from carelessness and fickleness is what we might call undue ambivalence. An epistemically responsible agent judiciously gathers and assesses reasons for and against a consideration. If the considerations on each side of a dispute are about equally balanced, there is no basis for favoring one over the other. If the situation requires a choice, the agent might as well flip a coin. If no decision need be made, she should suspend judgment, since, as far as she can tell, little favors one alternative over the other. Ambivalence then is appropriate. An unduly ambivalent agent takes reasons to be equally balanced when they are not. She may be inept at assessing the weight of reasons. Or, like the insecure student, she may lack confidence in what she is inclined to take the weight of reasons to be. Either way, she too frequently favors suspending judgment and may advise her interlocutors to do likewise.

A conscientious epistemic agent recognizes and attempts to live up to her epistemic obligations. She is appropriately scrupulous in amassing and assessing reasons. She is not obsessive though. She is aware of and attunes her behavior to reasonable standards of how scrupulous she needs to be. She checks the cookbook to see how long it will take to bake the cake but, absent specific reason for concern, does not worry that the recipe contains a misprint, that the cookbook incorporates misinformation designed to ruin dinner parties, or that her oven has suddenly gone undetectably haywire. She tries to reflectively endorse only considerations that are worthy of endorsement. In accepting something, she adds it to the store of considerations she is willing to use in cognitively serious inferences and actions. So she restricts her acceptance to considerations that she takes to be adequately supported. Since acceptance involves storage in memory for future retrieval, she is averse to fickleness. She regards as worthy of acceptance only such considerations as can be expected to stand up to further scrutiny. She is apt, of course, to be wrong occasionally. Evidence may quickly emerge that something she considered acceptable is unfounded. Then she gives it up. Doing so is not fickle, for the issue is not the longevity of a conviction but the sensitivity to the strength of the reasons that bear on its

Epistemic Engagement

acceptability. Although we may expect that a settled conviction will stand the test of time, every conviction is vulnerable to be overturned when new evidence emerges that calls it into question.

The conscientious agent is neither cavalier nor careless in her acceptance. She gives reasons their due. Unlike her unduly ambivalent counterpart, she does not set the threshold on acceptance at an unjustifiably high level. Having high standards may seem admirable. But setting too high a threshold on acceptance epistemically impoverishes, depriving the agent of resources for inference and action that she is epistemically entitled to. Conscientiousness alone does not determine where the threshold should be set. Rather, it reflects an attunement to an appropriate threshold. Conscientiousness thus involves being sensitive to both the standards for and the consequences of accepting, rejecting, or withholding a candidate commitment.

Some of the consequences involve the impact of acceptance on a community. As a member of an epistemic community, an agent is in a position to contribute to the store of resources that the community possesses. She might supply new information, new methods, new instruments, or new standards. She might reinforce or responsibly challenge commitments the community already holds. To make such contributions responsibly, she must be conscientious. She must ensure that her contributions, positive or negative, satisfy or properly challenge community standards. Moreover, she must be willing and able to voice her reservations when they are appropriate. To remain silent when she has something of value to offer is to fail in her responsibility as a member of her realm of epistemic ends (see Johnson 2018).

The emphasis on community standards might seem to put her under the sway of public opinion. The community sets the expectations; she is expected to live up to them. This worry overlooks the fact that she is, and thinks of herself as, a contributing member of the community. As a legislating member of that realm of ends, she participates in forming, validating, endorsing, enforcing, and, if need be, revising those standards. If she is privy to a consideration that calls the adequacy of current standards or their typical application into question, she is expected to bring it to the attention of the community. If she is right in her assessment, the community may recognize that the standards need to be revised.

Open-Mindedness

Human beings are neither omniscient nor infallible. My understanding is limited and no doubt contains mistakes. Nor am I unique in this regard. We are all epistemically deficient to one degree or another. Moreover, we know this about ourselves and each other. Each of us improves her epistemic lot by depending on others. Although we all have our limits, ignorance and error are not evenly distributed. Others understand things that I do not, and harbor correct opinions about matters where I am in error. If I can draw on them, I can compensate for some of my epistemic shortcomings.

If my informants are members of a realm of epistemic ends that I belong to, relying on them is relatively straightforward. When they impart information that I do not have, I can reasonably assume that the information satisfies our shared epistemic standards; when they correct my errors, I can reasonably assume that the basis for the correction lies in shared standards as well. There may be local squabbles about who is right, but even when there are, they take place on a relatively stable basis of agreement. A modicum of open-mindedness may be required for me to acknowledge the possibility that I am mistaken, or that the breadth of my understanding is not as great as I thought. But typically, having acknowledged my epistemic limitations, I can readily learn from my compatriots.

The difficulty comes when it behooves me to look beyond the boundaries of my particular realm of ends. Then open-mindedness is considerably more demanding. A behavioral geneticist maintains that a genetic trait causes human aggression; an ecologist maintains that environmental factors are the cause (see Longino 2013). The dispute implicates disagreements about methods, approaches, orientations, even definitions. For a geneticist to suspect that the ecologist might be onto something, she has to suspend or weaken a variety of commitments that constitute her understanding of aggression and how to investigate it, so that she can entertain an alien alternative. It is a stretch, but a potentially rewarding one.

Open-mindedness is an epistemic virtue that enables us to draw on the insights of others when they are alien to our own views. To the extent that an agent is open-minded, she has access to a wide range of potentially valuable epistemic resources. The difficulty is that to avail herself of these resources, she seems to have to suspend some of her own epistemic commitments. According to Riggs, "an open-minded person is characteristically (a) willing

Epistemic Engagement

and (within limits) able (b) to transcend a default cognitive standpoint (c) in order to take up or take seriously the merits of (d) a distinct cognitive standpoint, (e) and is sufficiently sensitive to cues indicating the existence of such alternative standpoints, (f) while having a well-calibrated propensity to exercise these abilities" (2019:150). A default cognitive standpoint is the network of relevant commitments that an agent reflectively endorses. What Riggs calls transcending that standpoint is simply holding it, or some part of it, in abeyance in order to entertain alternatives fairly and judiciously. In effect, the open-minded epistemic agent takes some of her commitments offline and entertains them and their alternatives, teasing out the consequences to see what they reveal. She retains higher-order commitments that enable her to critically assess alternatives.

Open-mindedness is not mindless tolerance. It consists in identifying alternatives worth entertaining, then judiciously assessing their strengths and weaknesses. If I want to extend my cognitive range and have reason to think that others have epistemic resources that I lack, I do well to try to access them. But I should only endorse them if they stand up to scrutiny. Open-mindedness is not gullibility.

Still, open-mindedness might seem unduly burdensome. If it required the agent to seriously entertain and assess every available alternative, inquiry would be hamstrung. Epidemiologists would be expected to devote their time repeatedly discrediting anti-vaxxers; astronomers, to repeatedly discrediting astrologers; evolutionary biologists, to repeatedly discrediting creationists. The worry is misguided. Open-mindedness does not require repeated reconsideration of positions that have already been reasonably rejected. Nor does it require devoting resources to assessing proposals that have nothing in their favor. Reconsideration must, of course, be possible. But reasons that justify reopening a dispute must be more than cosmetic repackaging of previously rejected arguments. Nor does the mere fact that some folks continue to endorse the discredited position constitute a valid reason to reopen a dispute. That lots of people still believe that the sun travels across the sky is no reason to rethink our allegiance to heliocentrism. Once a position has been responsibly rejected, the issue can be considered settled unless good reasons are adduced to show that the rejection was misguided. Such reasons may emerge from a reanalysis of previous arguments against a position, but only if the reanalysis reveals what in law would be called *reversible errors*—errors of sufficient magnitude that had they not been made, the verdict would likely have been different. Reasons to reconsider may also stem from the discovery of significant new evidence—evidence that raises doubts about the verdict. The practice of considering arguments settled unless such reasons emerge is the epistemic analog of the common law principle of stare decisis.

As our understanding of a topic grows, so does our understanding of the range of alternatives we need to take seriously. Arguably, at the dawn of inquiry, every hypothesis deserved a hearing. If they knew no better, perhaps it was plausible for Homeric epistemic agents to think that Zeus's anger triggers lightning strikes. But the advancement of understanding narrows the range of plausible alternatives. Just as we refine our standards of evidence for determining which hypotheses are acceptable, we refine our standards of evidence about which hypotheses or approaches are worth entertaining. We learn to identify the communities whose viewpoints might augment or correct ours. Helen Longino (2013) contrasts a range of approaches to the study of human aggression. These include molecular genetics, behavioral genetics, neurology, human ecology, and psychology. An open-minded practitioner in any of these fields might benefit from insights she could glean from the others. The findings of neurology might reasonably be expected to bear on psychology, for example. But she would not and should not seriously entertain the hypothesis that human aggression is caused by demonic possession. Nor should she be considered dogmatic or closed-minded for dismissing that hypothesis out of hand.

As I've described the burdens of open-mindedness, there may seem to be a relatively sharp divide between what it demands within a realm and what it demands across realms. This is misleading. Within a realm, I can suspend judgment about a particular proposition, method, standard, or norm, or about a range of them, while retaining and relying for support on the remainder of my network of commitments. My compatriots draw on them to show me how to extend or correct my views. In the cross-realm case, I suspend more. But even there, I don't suspend everything. The fundamental requirements of consistency and coherence hold across realms, as does the methodological principle that like cases be treated alike. Where they are relevant, mathematics, logic, and statistics provide shared framing principles as well. When we are open-minded about insights from adjacent realms, a variety of substantive and methodological commitments may be shared. The molecular geneticist is apt to share a considerable range of

Epistemic Engagement

commitments with the behavioral geneticist. If so, she will have, in Riggs's terms, a relatively easy time transcending her default standpoint to find a foothold she and the behavioral geneticist share.

Suspending some of the constraints of one's own practice enables one to identify and exploit resources that another discipline provides. Biologists in a bioengineering lab, hoping to devise an artificial blood vessel, bring to the project a rich understanding of how organic blood vessels function. They may have little expertise in fluid flow. They expand their purview when they suspend detailed information about how one should think about veins as living tissue with blood flowing through them and consider the issue mechanically—as a problem about how fluid flows through a vessel, about what happens at the boundary between the fluid and the vessel, about what facilitates or inhibits the flow (see Nersessian 2022). They learn how to think about the phenomena differently as a result of suspending some of their own commitments and provisionally adopting or at least seriously entertaining those of another discipline or practice.

My discussion might suggest that open-mindedness is a matter of being receptive to the expertise of others with suitably elevated academic profiles. Such elitism is misguided. Expertise can be displayed by folks who live in a particular area or speak a specific dialect or have a distinctive range of experiences. Inhabitants of a region may have vital local expertise about their surroundings (see Barrotta and Monduschi 2018). Indigenous expertise may be grounded in traditions that are not incorporated into Western science (see Massimi 2022:350–352). Sometimes we can vet credentials to discover whose views are worth taking seriously. Sometimes we have to look for other evidence. But the fundamental question is not what credentials they have but whether they have expertise that is both relevant enough and trustworthy enough to serve our epistemic ends.

Sincerity

In discussing competence and conscientiousness, I initially focused on an agent's epistemic relation to a particular subject matter—to her mastery of the topic and her appreciation of and responsiveness to the reasons for the considerations she accepts. I then showed how the relevant dispositions bear on her relations to other members of an epistemic community. In discussing sincerity, I reverse the order. I begin by focusing on sincerity as a communicative virtue and go on to consider its contributions to an individual agent's intellectual character.

Sincerity, as I construe it, is the antithesis of duplicitousness. It is consonant with irony, satire, metaphor, and sarcasm, so long as the ironic, satiric, metaphorical, or sarcastic character of the message is manifest. We should not think that Voltaire was insincere when, after cataloging a list of disasters in *Candide*, he concluded, manifestly ironically, that this is the best of all possible worlds.

Perhaps sincerity then is simply a matter of telling the truth, conceding that the truth need not be literal. As Williams (2002) argues, the relation between sincerity and truth telling is complicated. There are such things as honest mistakes. Someone who reports that the seminar meets on Tuesday, not having heard about the schedule change, may seem to speak sincerely, even though the seminar has been canceled. Perhaps she could be faulted for her failure to check, but her sincerity, we might think, ought not be impugned. She said what she genuinely believed. Although such a conception of sincerity fits with our intuitions, it is too thin, for in making her statement, she implicates that it satisfies the evidence maxim. She gave her audience to believe that the seminar meets on Tuesday and gave them to believe that she had sufficient reason to think that it meets on Tuesday. Inasmuch as satisfying the evidence maxim is a requirement for speaking sincerely, her assertion does not qualify as sincere. This conception of sincerity diverges from everyday usage, but the divergence is needed to underwrite the idea that sincerity is an epistemic (and not just a moral) virtue and insincerity is an epistemic (and not just a moral) vice. Honest mistakes need evince no epistemic failure. We reflect this in everyday life when we excuse someone for an error by saying, "You couldn't have known." The error and irresponsibility lie elsewhere. Insincerity is a matter of misleading one's audience about the content of one's statement or about the adequacy of one's evidence for it, either intentionally or through culpable negligence.

An utterance is misleading when, whether or not it is true, it conveys something false. Often this is a matter of implicature. Ben assures his dissertation supervisor that his work is on track by saying, "I haven't finished checking the footnotes." His statement is true, but the reason he hasn't finished checking the footnotes is that having not yet written a word, there are, as yet, no footnotes to check. He is guilty of insincerity if he intentionally gives

Epistemic Engagement

her reason to believe something false, even though the vehicle for conveying that belief is a truth. Still, not every misleading implicature is a manifestation of insincerity. As Williams says, a statement says one thing but implicates many (2002:100). Someone might sincerely, but inadvertently, give her audience reason to believe a falsehood through an unsuspected, misleading implicature. Suppose a university admissions officer believes that students succeed mainly through hard work. He might take a recommendation that praises an applicant's native ability but says nothing about his work ethic to implicate that the applicant is smart but lazy. This implicature is available even if the recommender intended only to convey his admiration for the applicant's intelligence. Sincerity alone cannot preclude all misinterpretations. But it should block obviously misleading ones.

The issue is further complicated because not all communicative vehicles are propositional. Where they are not, they are not truth apt. A map, diagram, gesture, or picture can be accurate or inaccurate, but truth is not an appropriate metric. Still, such a representation may be misleading or obscure. A diagram can be defective by including too much information, not enough information, the wrong information, the wrong grain of information, or the wrong orientation toward the information for the purposes for which it is to be used (see Tufte 2001). It may implicate through what it displays, how it displays, or what it omits. This can be intentional or unintentional.

It might seem that since a diagram makes no statement, it cannot be either sincere or insincere, honest or dishonest. The critical question is how it is used. But the same might be said of a sentence. If a sentence is used in a way that purports to supply information, then it can be faulted if the information is irrelevant, inaccurate, or unfounded. If it is uttered as an exercise in elocution, such issues do not arise. If the diagram is proffered as supplying information about, say, variations in the annual rainfall in Peru, then it can be faulted as inaccurate if it misrepresents the rainfall. If the city map omits streets, then it misrepresents the layout, misleading users about how to get to their destination. If such representations are just used for decoration, their inaccuracy is irrelevant. For our purposes, the important point is that such inaccuracy can be intentional. If a nonverbal symbol is intentionally designed or used to mislead, its design or deployment involves insincerity.

Sincerity seems other-directed. Insincerity seems to be a matter of intentionally or negligently imparting information to others, even though one considers that information untrustworthy. Since we do not impart information to ourselves, it might seem that sincerity gets no first-personal foothold. This is not quite right. A person can be self-deceptive. If she is, she accepts a consideration against her better judgment. That is, she accepts it, despite realizing or suspecting, or being capable of realizing or suspecting, that it violates her own criteria for reflective endorsement. How this is psychologically possible is a matter of dispute. But it is possible, and it is epistemically damaging. It might be due to problematic compartmentalization. It might be due to carelessness, sloppiness, or wishful or fearful thinking. It might, as we saw, be due to her failure to trust her own judgment. Whatever the reason, the self-deceptive agent is not honest with herself.

Epistemic Empathy

A message may be competent, conscientious, sincere, and the fruit of open-minded inquiry yet still be abortive, for it might be couched in a mode of expression that fails to align with the audience's needs. It might, for example, be epistemically inaccessible to the audience, perhaps being cast in unfamiliar technical jargon. It might adopt an inappropriate orientation—perhaps taking a third-person perspective when a first-person perspective is called for. It might convey reams of irrelevant information, obscuring the points of interest to the audience. A good informant needs to be more than competent, conscientious, and sincere. She needs to be attuned to the communicative situation and sensitive to its demands (see Gardiner 2022). She should make her contribution intelligible, apt, relevant, appropriately oriented, and true enough so that the message it conveys is epistemically useful to her intended or expected audience. A non-accidentally good informant displays what we might call "epistemic empathy" (see Jäger and Malfatti 2020; Elgin 2020). She adopts her audience's orientation in order to grasp their interests and abilities so that she can tailor her responses accordingly. This is not to say that she always accepts what they accept. Rather, epistemic empathy enables her to see how things look from their perspective. This puts her in a position to frame her offering in a way that her audience can understand and benefit from. She appreciates what and why they seek to understand, and she recognizes what epistemic resources she can draw on—their presuppositions, background information, default orientations. This enables her to choose a vocabulary, grain, and perspective that will make the information she imparts intelligible and useful to them. A good informant, moreover, recognizes and accommodates impediments that are apt to interfere with her audience's ability to access or deploy the information she provides. If she has reason to suspect that they lack mathematical sophistication, she will, as far as possible, omit equations. If she has reason to suspect that they are well read, she may invoke literary allusions. If she has reason to think they are sports buffs, she may deploy sports metaphors. She may decide that for a given audience, it is best to present the material literally or figuratively, mathematically or narratively, verbally or diagrammatically. Her choice depends on both the material to be presented and the resources that the audience can draw on to grasp that material. If she appreciates her audience's expectations and resources, she is sensitized to specific possibilities of ill-advised implicatures that her statements might otherwise invite. She is then in a position to block these implicatures. By adopting her audience's perspective, she can see what resources it provides, what holes she needs to fill, and what pitfalls she should take pains to avoid. Adopting her audience's perspective thus enables a good informant to choose a vocabulary, orientation, and mode of presentation that suits current communicative needs.

It might seem that such attunement is possible only when information is targeted to a specific audience. Len knows quite well what his colleagues bring to the table—they all have advanced degrees in biochemistry. In making his presentation, he can rely on their shared knowledge. But much information is impersonally shared. In publishing an article or writing a blog, it might seem, I address an audience of strangers. I do not know who they are or what epistemic resources and expectations they bring to their reading of my work. I lack the information I would need to attune my message properly. This worry is unduly pessimistic. Granted, I don't know the details of my readers' backgrounds in the way that I know the details of my close colleagues. I cannot draw on the fact that we went to the same schools or read the same books. So I ought not presuppose specific, detailed information that is not widely shared. But even when writing for a general audience, we make assumptions about them—their level of education, their vocabulary, their interests, common information that we can draw on. We regularly and automatically key our verbal presentations to our expected audience. Appealing to common ground is standard, responsible communicative behavior. That it involves epistemic empathy is often overlooked.

A good informant's contributions then are not merely accurate; they are relevant, apt, and useful to their audience. Such an informant is not merely reliable; she is trustworthy. The difference is this: reliability is statistical; trustworthiness is normative. A trustworthy informant is epistemically responsible. She can typically tell whether she has the information the inquirer seeks. This, as we have seen, involves self-knowledge as well as knowledge about the topic. She voices an opinion only if she has good reason to think that she is competent. She calibrates her confidence to her competence. If she lacks confidence in her opinion, she makes that clear. She is sincere. She ventures information only if she thinks it is accurate enough and adequately backed by evidence. She is respectful of the standards of accuracy that are appropriate in the circumstances. She is, moreover, socially sensitive. She conveys information in a way that is attuned to her inquirer's needs, interests, and abilities. She will not, for example, give a highly technical, jargon-ridden, virtually unintelligible answer to a layman's question. Nor will she talk down to him. She will not adopt an orientation on the problem that is irrelevant to his needs. She will not be unduly precise or unduly imprecise. And so on. That is, she conforms to Gricean maxims.

A merely reliable response to a request for information could easily be unintelligible, irrelevant, wrongly calibrated, or otherwise unsuited to the inquirer's needs. A chatbot could, like a barometer, provide an accurate answer to the question whether charged leptons undergo strong interactions. But if it was not sensitive to the interests, background beliefs, and abilities of its audience, its answer might be useless. A good informant must be suitably sensitive to the topic and to her audience. These requirements are normative. Ceteris paribus, an informant is blameworthy if she lacks them. This is not to say that she is blameworthy *for* lacking them. Rather, ceteris paribus, she is blameworthy if she informs although she lacks them. The norms are, I suggest, at once ethical and epistemic, for the information is presumptively trustworthy if the source is.[4]

Conclusion

My focus in this chapter has been on a cluster of epistemic virtues that figure in wielding information responsibly—that is, on using or imparting information as a basis for inference or action—when one's ends are cognitive. These virtues underwrite trustworthy communication. But trustworthiness enters the picture earlier. It figures in the generation of information that is worthy

Epistemic Engagement

of trust—in the calibration of instruments, in the design and execution of experiments, in the definition of measures and the design, manufacture, and use of measuring devices. It figures in the selection and use of methods of inquiry. In all these cases, competence, conscientiousness, open-mindedness, and sincerity are critical. Unless information generation is trustworthy, the downstream uses—however competent, conscientious, open-minded, and sincere—are at risk. Trustworthiness thus lies at the heart of inquiry.

4 Epistemic Dynamics

Introduction

Human beings are fallible, finite, and capable of epistemic self-improvement. In our efforts at self-improvement, we come to grips with our fallibility and finitude. We recognize that there is much that we do not know, much that we do not understand. We acknowledge that some of what we accept is apt to be mistaken. Minimally, we devise epistemic methods and practices that allow for correction, refinement, and expansion of our current take on things. Preferably, we devise methods and practices that foster (rather than merely allow for) correction, refinement, and expansion.

Methodological Common Ground

Whatever the criterion for epistemic acceptability, our best prospect of arriving at acceptable conclusions is the method of reflective equilibrium. We begin with whatever opinions, methods, mechanisms, and norms we take to be viable. We augment and revise them when they seem inadequate—perhaps because they are jointly inconsistent, mutually untenable, collectively implausible, or simply gappy—or when we think we can do better. Through a process of adjudication, we bring considered judgments about a topic and the appropriate ways of assessing them into accord. When a network of commitments consists of elements that are reasonable in light of one another and the network as a whole is as reasonable as any available alternative in light of the relevant antecedent commitments, they are in reflective equilibrium. They are, at the moment, the best we can do. Is that enough? Is a network of cognitive commitments epistemically acceptable simply because it is the product of the method?

82   Chapter 4

Some say no. They may concede that the method of reflective equilibrium is the most practicable method for arriving at knowledge or understanding. To start with anything other than our best guesses about how things stand in a domain, how to investigate that domain, and how to evaluate our findings seems mad. But even if the method is epistemically optimal, it does not follow that being in reflective equilibrium constitutes being epistemically acceptable. Perhaps being the product of such a procedure is indicative, but not constitutive, of acceptability.

It is widely held that the criterion for knowledge is non-fortuitously justified true belief. A belief satisfies the non-fortuitousness requirement when its truth maker and its justification properly align. Then, given its justification, its truth is no accident. Let us call this the traditional epistemological criterion for knowledge.[1] Some epistemologists maintain that understanding is a special sort of knowledge—for example, knowledge of causes or mechanisms or dependence relations (see Grimm 2014; Khalifa 2017). If so, the traditional criterion applies to understanding as well. On a traditional epistemological view, a network of considered commitments in reflective equilibrium might nevertheless be epistemically unacceptable. Perhaps the network contains falsehoods or truths that lack adequate justification. Perhaps luck plays too great a role or the wrong kind of role in the fact that some of the truths it contains are justified. These are familiar epistemic concerns.

In any given case, we can improve our epistemic prospects by checking for errors, strengthening justification, reinforcing connections between justifiers and truth makers. In so doing, we generate a broader, more tightly woven tapestry of commitments. It is clearly an improvement. But the disquieting gap between reflective equilibrium and the requirements embodied in the traditional criterion remains. There's no guarantee that our best methods are up to the task that traditional epistemology sets for them.

The typical accommodation to this predicament is to set a threshold for acceptability. Even if we cannot be certain that the truth of our beliefs is non-fortuitously tied to their justification, we can reach a threshold where the reasons supporting our beliefs are good enough. When the threshold is reached, the justification is sufficiently strong, and the alignment of justification conditions and truth conditions is sufficiently close that further worries can be set aside. That being so, it is held, if a belief that meets the threshold is true, we know.

One drawback to such epistemic accommodationism is that the threshold seems somewhat arbitrary. Why not a bit higher or a bit lower? Perhaps it would be better to let the threshold vary with context. Another worry concerns the basis for setting the threshold. Maybe it is reached when justification is sufficient for testimony (Hannon 2019) or when it is sufficient to close inquiry (Rysiew 2012) or when the probability of a contention, given the evidence, reaches a specified level. I won't pursue this issue. Rather, I will argue that the traditional epistemological stance impedes epistemic progress; taking reflective equilibrium as *constitutive* of acceptability fosters it.

The impediment is this: regardless of how or where the threshold is set, when a consideration reaches that threshold, traditional epistemology considers the matter closed. We are in a position to know that p and to understand that or why p caused q. There is no incentive to inquire further. Any remaining doubts about the matter are suppressed or dismissed as idle if not neurotic. Epistemic agents are fully entitled to use considerations that meet the threshold as a basis for inference and action. Such epistemic security seems welcome. If p is in fact a non-fortuitously justified true belief, and if all one wants to know is whether p, then at least when the stakes are not exorbitant, perhaps reaching the threshold is enough.

Still, the grounds for accepting p or for accepting that p caused q—the reasons to think the judgment met the specified threshold—are that it is suitably integrated into a network in reflective equilibrium. That should give adherents of the traditional stance pause. Integration into such a network shows only that p or that p caused q is reasonable in light of a variety of other commitments that are collectively reasonable in the epistemic circumstances. This is consonant with the network's incorporating falsehoods, measurements with significant error bars, modes of inference that are less than fully reliable. Some, such as those I have characterized as felicitous falsehoods, are epistemic assets (see Elgin 2017:23–32). Others are, and may be recognized as, deficiencies that are consonant with the system as a whole being as good as any available alternative. They are, we may think, shortfalls we can live with, since they are not too serious, and any available alternative is at least as unsatisfactory. Perhaps we settle for an approximation, since an exact solution eludes us and the approximation is not off by much, at least in the range we care about. Perhaps we reluctantly rely on a test with a few false positives, since we have nothing better to go on and anyway there aren't too many of them. This may be entirely reasonable in the epistemic circumstances. Nevertheless we might and arguably should be reluctant to consider the verdicts yielded by such a system to be completely settled.

By taking reflective equilibrium to be constitutive of adequacy, we convert this seeming liability into an asset. Rather than maintaining that a network in reflective equilibrium is acceptable come what may, we consider it merely as acceptable as any available alternative here and now, and we treat it as a good enough platform to build on. Although probably somewhat flawed and definitely incomplete, it enables and incentivizes us to leverage understanding and to protect ourselves against the perils of misunderstanding. We treat current understanding, however good it is, as a way station on the route toward better understanding.[2]

Traditional epistemology focuses on propositions: traditional theories of knowledge on individual truth-value bearers; traditional theories of understanding (such as those that hold that understanding consists of knowledge of causes or dependence relations) on constellations of truth-value bearers. Although the methods, measurements, and standards of justification that underlie our conclusions are implicated in the acceptability of both individual propositions and networks of propositions, traditional epistemology tends to bracket them—to take them for granted rather than acknowledging and assessing their role. Still, in holding that p is justified, we commit ourselves to p's being underwritten by appropriate methods, standards, and measures.

Since an empirical claim is justified only if the evidence supports it, its being justified depends on the adequacy of the methods used to generate and validate the evidence. One issue is what qualifies something to serve as evidence.[3] Even the crudest inductive cases are more complicated than they seem. Observational evidence supports the conclusion that all ravens are black only because we have been careful to gather that evidence under conditions where (we think) the lighting is not deceptive and our perceptual faculties are not impaired. We restrict the observations that count as evidence to those made by people who (we think) can reliably tell ravens from other birds. We make observations in circumstances where (we think) variation in plumage would occur if there were any to be found. We exclude from our database sightings where (we think) these conditions may have been compromised. In less crude cases, we process raw material to obtain data that qualifies as evidence. We filter out (what we consider to be) impurities, control for (what we think might be) confounds, analyze complexes to identify and isolate

Epistemic Dynamics

their components. Generating and gathering evidence may involve instrumentation. Then the acceptability of our evidence depends on the accuracy of our devices—eyeglasses, thermometers, mass spectrometers, magnetic resonance imaging (MRI) scanners—and our understanding of their powers and limitations. Statistical methods often play a prominent role in vindicating the claim that a body of evidence is creditable. They dictate the requisite sample sizes and distribution. They determine what sort of information it is appropriate to extract from a database. Inferential methods—some of them non-truth preserving—figure in the paths from evidence to conclusions. We depend on models, idealizations, simplifications, and analogies that depart from truth, just as we do on mathematical and formal logical reasoning that preserves truth.

Similar points apply to measurement. Evidence derived from measurements depends for its cogency on the accuracy and adequacy of our measuring techniques and devices. If a device is inaccurate in cold temperatures or in strong magnetic fields, the readings obtained in such circumstances are potentially misleading. If the orientation or focus of the detector is out of kilter, its results are not to be trusted. If a phenomenon is affected by being measured, then the result of the measurement does not directly disclose its natural, unmeasured state. If a device too readily loses its calibration, our confidence in the readings it generates should wane.

I've put these points in terms of the methods and measuring devices used in the natural sciences, but they hold equally for social scientific methods and measurements—focus groups, ethnographic studies, and public opinion polls, for example. The contention that the results of such investigations are characteristic of the actual opinions and normal behavior of the populations being studied is subject to challenge. The results may be skewed by the way they are studied or even by the fact that they are studied. The same concerns also hold in mundane epistemic settings. We rely on gas gauges, GPS devices, casual observations of our fellows, and all manner of apps. If they are ill-suited to the conditions under which we use them, the conclusions we draw are epistemically unfounded.

Epistemic norms, standards, and criteria play crucial parts as well. Requirements on the variety of evidence, the accuracy and precision of measurements, the criteria of relevance, the appropriateness of orientation and grain underwrite evidence and vindicate results. The conclusion that all ravens are black is unacceptable if observers only looked at the ravens in one neighborhood or only looked at black-and-white photographs of ravens or restricted their evidence to observations of non-black non-ravens. What counts as evidence, what counts as misleading evidence, and what counts as sufficient non-misleading evidence are implicated in our judgment that a given proposition or network of propositions is justified by the evidence.

It might seem that epistemology could acknowledge such dependencies but bracket them, arguing that standards are the province of epistemological metatheory and that measures and methods belong to the context of discovery rather than the context of justification. If we want to know whether a conclusion is acceptable, we need to know whether (and perhaps how) it is justified. But the path by which we arrive at that conclusion is a matter for the psychology or sociology of inquiry. Tanja Rechnitzer (2022) argues persuasively that reflective equilibrium is a viable method in the context of discovery. Methods, standards, and measures play a vital role there. I think that they are equally vital in the context of justification or, more accurately, that the context of discovery and the context of justification cannot be prized apart. The methods and measures I have been talking about are methods and measures that figure in the justification of the conclusions we reach. A consideration's satisfying a given standard constitutes its being justified only if the standard itself is apt. Even if the standard's justification derives from metatheory, its aptness depends on its suitability to the circumstances where it is used. Only if the methods, measures, and standards are trustworthy and appropriate are the conclusions they support acceptable. Justification is a product of justifying. How we go about justifying a conclusion determines whether what we produce is a justification for it (see van Fraassen 2008).

This does not give the methods, standards, and/or measures epistemic priority. What vindicates specific methods, standards, and measures is that they figure in networks that contain and sustain propositions we consider justified. What vindicates propositional commitments is that they are underwritten by methods and measurements that we consider trustworthy and that satisfy standards that we consider sound. There is an intricate dialectical interplay among norms, methods, and judgments about matters of fact. Each depends on its mesh with the others. A method or measurement is acceptable only if the verdicts it supports are justified. A norm, criterion, or standard is acceptable only if its satisfaction yields verdicts that strike us as right (see Goodman 1983:63–65). Although often consigned to the periphery, methods, measures, and standards are implicated in the justification of

Epistemic Dynamics

factual claims. Traditional epistemology is committed to epistemic holism, whether wittingly or not.

Advancing Understanding

If we reject the idea that inquiry terminates when equilibrium is reached and reject the idea that methods, measures, and standards simply scaffold understanding without being integral to it, we can exploit the resources holism provides to advance understanding. Since acceptability involves a network of commitments in reflective equilibrium, a survey of the network should enable us to discern the various components and the ways they depend on and support one another. We thus have a perspective from which to identify vulnerabilities and opportunities. It follows from fallibilism that some of the considerations we justifiably accept might nevertheless be flawed. A contention that is supported by what we justifiably consider sufficient evidence could still be false; a method that works in a restricted domain might fail in a broader domain. A standard that seems viable at one grain may prove problematic at a different one.

The process that generates reflective equilibrium affords opportunities to correct errors and shore up weaknesses. An initially tenable commitment, however plausible in isolation, should be amended or rejected if it lacks support from or fails to lend support to other elements of the system. When we start the process with separate, individual commitments, correction and amendment seem straightforward. We fine-tune the separate commitments until they mesh. But once a system achieves equilibrium, it might seem, this advantage vanishes, for if the system as a whole is the bearer of acceptability, how will we know what to reject or amend? If we remain dissatisfied, we could of course reject the whole thing and start over from scratch, but that would sacrifice the epistemic progress embodied in the system. Holism seems epistemically cumbersome.

But holism is consonant with revisability. Quine (1980) maintains that any commitment can be held true come what may, so long as we make compensatory revisions elsewhere in the system. This might seem to suggest that holism allows us to revise any commitment we please, in any way that we like. If so, we are free to be cavalier or quixotic in making revisions. That would be unduly permissive. But criticisms of Quine's criterion overlook the force of "so long as we make compensatory revisions elsewhere in the system." Holism has a very strong coherence requirement. If we revise or reject a given element, a variety of other elements also need to be modified. If we reject a judgment that is based on evidence, we have to either reject some of the evidence or reject the standard that deems that evidence adequate. That will require rejecting the evidence for the evidence, the other commitments that are based on that evidence, the other commitments that are vindicated by that standard, and so on.

Revisions ramify. Quine favors revisions at what he calls the periphery. These are commitments that are minimally integrated into the system. Because they have few and weak connections to other elements of the system, their revision or rejection is not costly. To be sure, it is open to us to revise the law of noncontradiction or the conviction that dogs are domesticated or the use of thermometers to measure temperature if we are willing to pay the price. But each of these is so tightly and intricately interwoven into our understanding that the price is likely to be high. It may be that any of these has to be given up, but he recommends that we consider less costly changes first.

Quine's discussion shows that although every element of a system is in principle revisable, they are not all equally open to revision or rejection. Different commitments have different measures of epistemic inertia. A peripheral commitment—one that is loosely interwoven into the network—can be given up rather easily. We don't lose much, and we don't have to revise much if we recognize that, for example, a particular observation was probably flawed or that there was likely some sort of undetected interference in a single measurement. Still, it won't do to dismiss even these cavalierly. Before we dismiss an odd reading as due to undetected interference affecting the measuring device, we do well to take other measurements under the same circumstances, using distinct devices of the same kind. If they give more plausible readings, we may be safe in our verdict. But if they also give discrepant readings, then we should look more deeply into the connection between the device and the measurements we expect it to yield. Are we sure that the device is faulty? Or should we entertain the hypothesis that the phenomenon it measures is behaving in a way we had not anticipated or that the interaction between the measuring device and the phenomenon it seeks to measure is not what we thought? Even if we conclude that the device malfunctioned, it might be a good idea to figure out why. Before we disregard an observation, we should look more carefully—perhaps simply

Epistemic Dynamics

by looking again, perhaps by adopting other perspectives, perhaps by asking others what they see—to be sure that it was in fact discrepant.

The discussion so far might seem to follow Quine in favoring a principle of minimal mutilation (1990:14–15). In making revisions, he thinks, we should preserve as much of the current system as we can. This is not always the best strategy. Rather, we should preserve what we consider most worthy of preservation. The ramifications of revising a central commitment may be epistemically beneficial.

Here is an example: A central commitment of Michael Strevens's *Depth* (2008) is that all scientific understanding is causal. Suppose we start with his system. A network that satisfies his requirements restricts the justification for a scientific claim to the purely logical or mechanistic: p justifies q only if p either logically entails q or p constitutes a mechanistic explanation of q. The commitment to causal mechanisms, he maintains, is central to acceptability in natural science. This excludes or at least downplays the epistemic significance of nonmechanistic models.

The Lotka–Volterra model of predator–prey relations prescinds from mechanisms. It displays the interdependence of the sizes of predator and prey populations without any attempt to explain how the two populations modulate their size and without any suggestion that all population pairs do so in the same way. According to Strevens (2017), the model is not explanatory. But so long as certain constraints are satisfied (which they often are), just how the predator–prey species modulate their populations seems irrelevant. The decrease in available prey explains the decline in the predator population, which in turn explains why the prey population rebounds, and the cycle continues. The effectiveness of the model suggests that science should not limit itself to mechanistic explanations.

It might seem that all that this shows is that if we relax our standards on understanding, more things count as being understood. It is rather like saying that if we lower the passing grade, more students will pass. That's hardly a surprise. But the point here is not that more things count as being understood if we relax the commitment to mechanisms. It is that *a pattern emerges* that is indifferent to mechanisms. If we insisted on a mechanistic account, that pattern would be invisible. We would not be able to see how fox–rabbit population pairs are like starfish–mollusk population pairs and like loan shark–indigent borrower population pairs. At a certain level of abstraction, differences in reproductive mechanisms don't matter. If we start with Strevens's network, considerable revision is necessary to accommodate models like Lotka–Volterra. But the payoff is great enough to make it reasonable to revoke the restriction to causal mechanisms and open the door to a broader range of explanatory factors. In surveying an epistemic network, we focus on relations of support. This enables us to entertain the consequences of relaxing or strengthening requirements, of suspending or introducing commitments.

If acceptance amounted to holding that the propositional commitments of a network were true and the methods that justified them were fail-safe, the flexibility I advocate would be epistemically risky. But if we recognize that an accepted system owes its acceptability to its being the best we can currently do, the situation is different. Then acceptability invites rather than resists revision. It encourages and equips us to look past boundaries that may turn out to be unduly restrictive. There is nothing easy or automatic about such extrapolations. Nor is success guaranteed. But because connections among commitments are flexible, because the parameters in an accepted theory are regarded as revisable, they can be relaxed or tightened. And because they can, questions about the benefits and costs of relaxing or tightening them can and often should be entertained.

Dynamic Holism

The picture I have sketched so far is of a network in reflective equilibrium, which is open to inspection. It is laid out before us. If we find an opportunity to extend the network or reinforce the elements in it, we do so. If we find an error or a vulnerability to error, we do what we can to correct it, making revisions that preserve or readily restore equilibrium. This is right as far as it goes, but it is misleading. It still seems to treat a system in reflective equilibrium as a finished product. What this characterization conceals (or anyway fails to emphasize) is that one of our cognitive ends is improving the understanding that the system already affords. An epistemic system in reflective equilibrium should equip agents to identify and correct errors and to expand its range. That is, it should enable leveraging of the understanding it provides.

Surveying a system in reflective equilibrium reveals the warp and weft of the network of commitments. This enables us to understand how we understand what we understand about the phenomena. Such a survey reveals dependence relations among our various commitments. It shows how the conclusions we accept are beholden to methods, mechanisms, and standards for support. It thus reveals vulnerabilities in what we accept and opportunities for improvement. A network's equilibrium is unstable if the methods or mechanisms involved are unacceptably sparse, limited, precarious, or flawed. For example, many psychological studies are done on undergraduates—primarily on students in large introductory psychology courses. Perhaps the evidence gained from such samples of convenience is good enough. Maybe undergraduate psychology students in an American university are sufficiently like other human beings in their reactions to a phenomenon like cognitive dissonance that general conclusions can be drawn. Still, on reflection, we should prefer a broader, more variegated, more representative sample. A device that has been tested in a limited range but used to generate measurements beyond that range might also give us pause. A thermometer that yields consistent, plausible readings between 0°C and 100°C might be useless or untrustworthy beyond that range (see Chang 2004:152–158). Confidence that its readings are generally reliable might seem unfounded.

A survey of a network may disclose that standards are too broad, too narrow, or ill-aligned with the phenomena they purport to bear on. If, for example, we restrict acceptability in empirical domains to the results of double-blind randomized controlled trials, some well-founded conclusions have to be set aside. There are, for example, no such trials investigating whether dental flossing is beneficial. It would be morally irresponsible to ask people to be in the control group. Nor could such a study be double blind, since the subjects would know whether they floss regularly. Still, dentists and hygienists can tell which patients floss regularly and can show that failure to floss has manifest, deleterious dental consequences (Oreskes 2019:118–127). If we broaden the standards of acceptability and count observations of dentists and hygienists as evidence, the hypothesis that flossing is beneficial can be epistemically justified.

Leveraging Understanding

Rather than just correcting errors when we encounter them and extending our range when opportunities suggest themselves, it would be better to design our systems to leverage understanding. In finance, leveraging consists in using borrowed capital as a funding source to increase one's asset base. If the rate of return on an investment is greater than the interest on the borrowed capital, the investor makes a profit. She can then use that profit as collateral to borrow more money and leverage her asset base further. This is a familiar and effective investment strategy. Something similar holds in epistemic contexts. Thinkers, as it were, "borrow" considerations that are less than adequately justified and tentatively use them in cognitively serious inferences and actions. If those inferences and actions advance understanding, the suspect considerations are vindicated. They retroactively generate their own support. This sounds question-begging, but I will argue it is not.

Suppose a network of epistemic commitments that seems pretty good is not entirely satisfactory. If it is not in reflective equilibrium, the path forward is clear: revise commitments to bring them into accord, not being content until they are reasonable in light of one another and the network as a whole is as reasonable as any available alternative. It might seem then that our problem is solved. But even if a constellation of commitments is in reflective equilibrium, it may strike us as inadequate. Reflective equilibrium assures that it is at least as good as any available alternative. Sometimes that bar seems too low. Perhaps commitments in the network lack precision. Perhaps they are qualitative when quantitative measures would be preferable. Perhaps currently undefeated considerations undermine confidence. We may suspect that there are misleaders but that our current resources do not enable us to identify them. Even though current standards are met, the network still looks a bit dodgy. Perhaps the network raises questions that it cannot answer or opens up vistas for exploration without providing roadmaps indicating how to proceed. Although the current network of commitments is acceptable as far as it goes, it does not, we may think, go far enough. Being epistemically greedy, we want more.

Consider a couple of cases. Fred and George disagree about the weather. Fred maintains that yesterday was warmer than today; George insists that it was cooler. Our current network may be tolerant of divergence within the range where they disagree, recognizing that the boundaries between warm and cool are vague and somewhat subjective. Still, if we want to use their reports as indications of trends in the weather, we are apt to find the situation epistemically unpalatable. To take another case: Some measurements rely on imperfect proxies. We cannot directly measure the magnitude we want to measure, but we can measure something else that we have reason to think correlates with that magnitude. Oxygen isotopes in ice cores correlate, albeit imperfectly, with temperature. Because the correlation is imperfect, taking

Epistemic Dynamics

them to be a measure in assessments of climate change, even if this is the best we can do, may seem epistemically dicey. These are familiar predicaments. Available considerations strike us as inadequate for current epistemic purposes. The question is how to cope—how to responsibly build on the level of understanding we have already achieved.

The safest strategy is to suspend judgment until such matters can be resolved: simply take no stand on the issues in dispute. This minimizes the danger of accepting errors. But in suspending judgment we restrict our epistemic resources. That may be costly, as it results in our having fewer, and perhaps less fruitful, considerations to draw on in our efforts to further our epistemic objectives. A more daring strategy is to accept a consideration that does not satisfy the current standards of acceptability but do so with reservations. This would augment our epistemic asset base. It would also increase our vulnerability. Accepting a doubtful consideration threatens to inadvertently introduce error. We owe it to ourselves and our community of inquiry to discharge the debt.

To accept a consideration is to be willing to use it in inference or action when our ends are cognitive (Elgin 2017:19). To accept it with reservations is to incorporate sensitivity to its precariousness. For such a strategy to be reasonable and responsible, it must satisfy the standards of the epistemic community. That, in turn, requires that the community have relevant standards. If it does, then like lending institutions that back leveraged financial transactions, the community has criteria for what risks it deems worth taking, given the rewards we reasonably hope to gain. If a community of inquiry deems epistemic leveraging valuable, it should develop and deploy such standards. One could, of course, go rogue and accept a consideration with reservations whether or not the community sanctioned doing so. But the availability of and conformity with suitable standards makes doing so epistemically responsible.

Accepting with reservations is more common than my discussion might suggest. Climate scientists take the isotope ^{18}O as a proxy for temperature and treat different levels of ^{18}O in ice cores as proxies for past differences in temperature. They do so even though they acknowledge that the correlation between ^{18}O and temperature is imperfect. In regarding it as good enough in relevant respects for inquiry to proceed on the basis of it, they hold that investigations that treat ^{18}O as a measure of temperature will be more fruitful than investigations of the same phenomena that reject or withhold acceptance of the proxy. Inasmuch as they are trying to ascertain changes in temperature over eons and have no direct evidence to draw on, they need to use proxies. Since no proxy is perfect, their only way forward is to accept measures afforded by imperfect proxies. As far as possible, they hedge their bets. Where there is overlap between the ranges of different proxies for temperature, they calibrate them against each other. Thus, for example, they test the results of ice-core readings against the results of tree-ring readings. Agreement among them strengthens the tenability of each. To accept such proxy results without reservations would be irresponsible. It is preferable to be aware of the epistemic risks being taken and, as far as possible, to hedge our bets.

Previously, I argued that to achieve reflective equilibrium, a network can incorporate considerations that have no independent warrant (see Elgin 2017:65–66). Evolutionary biology is entitled to a commitment to transitional species to bridge gaps in the fossil record. Although there is no independent evidence for them, biologists have, and should have, no reservations about accepting them. An understanding of evolution that incorporates transitional species is more reasonable than one that does not. The cases that concern me here are different. The problem is not just that, apart from their contribution to reflective equilibrium, there is nothing to be said *for* them. It is that, although they may contribute to reflective equilibrium, there remains something to be said *against* them. Outright acceptance seems unwarranted.

Although my examples focus on the issue of accepting a single doubtful commitment, epistemic leveraging is not restricted to such cases. We may accept a network of commitments with reservations, recognizing that, as currently configured, it cannot answer seemingly legitimate questions or solve salient problems it confronts. Perhaps it sharply differentiates seemingly similar cases without suitable motivation for drawing the lines where it does. Perhaps it does not extend as far as we think it should but provides no justification for the boundaries it sets. Perhaps the connections among the various commitments seem slightly spurious or just too loose. Despite its weaknesses, we may consider it good enough to serve as a basis for further inquiry. In accepting it, we anticipate that inquiry will discharge the epistemic debt by showing where and how it was correct to the extent that it was, and perhaps why it countenanced the misleaders or was saddled with the other inadequacies that it displayed.

Epistemic Dynamics   95

As in finance, leveraging in epistemic contexts is Janus faced. The borrower should appreciate both the risk he undertakes and the rewards it promises. Borrowed epistemic capital consists in less than adequately supported factual, normative, or methodological commitments. To epistemically borrow a commitment is to treat it as acceptable—that is, to let it function in inference or action as though it were acceptable—in the face of non-neurotic doubts about its acceptability. To vindicate doing so is to show that it figures in the advancement of understanding. To leverage an epistemic investment is to design the resulting network in such a way that it can serve, perhaps with reservations, as an acceptable base for further improvements.

Epistemic Iteration

Hasok Chang's *Inventing Temperature* describes the co-evolution of the thermometer and temperature—the magnitude the thermometer measures (2004). Here I present a highly schematic review of part of the development Chang describes. My goal is to highlight one way that agents make epistemic progress by leveraging networks of commitments that they know to be inadequate. The commitments may be inadequate because they contain infelicitous falsehoods or because the truths they contain or the methods they rely on are ill-suited to current epistemic ends. Whatever the reason for their inadequacy, they do not as they stand serve our epistemic purposes. Chang calls such leveraging "epistemic iteration." He explains, "In the process of epistemic iteration, we knowingly start inquiry on the basis of an imperfect starting point, and use the outcome of that inquiry in order to improve its own starting point" (2022:239). This sounds not only question-begging but also self-defeating. *Inventing Temperature* shows that it is not.

Long before the invention of the thermometer, people recognized that they were sensitive to differences in warmth.[4] They knew, for example, that winters are colder than summers and that water freezes when the weather is sufficiently cold, melts when the weather gets warmer, boils when heated over a flame. They knew that although it was chilly enough to want a jacket at dawn, by noon the weather might warm up enough that they could comfortably do without one. Because they recognized that many of these phenomena exhibit regular patterns that are not wholly subjective, they took it that phenomenal states of warmth track something. And because at least some of the patterns seem important, they thought that it would be epistemically worthwhile to learn what they track.

They also recognized that subjective assessments of warmth are sometimes at odds. The wind feels cool to one person, mild to another. The hallway feels warm when one enters from outdoors cool when one enters from the cozy kitchen. The lake seems chilly when one first dives in mild a few minutes later. Nor could they dismiss variations in verdicts on the grounds that they simply reflect variations in the reactions of different people or at different times. It is possible to generate the troubling inconsistency in a single individual's phenomenal state at a single moment. Place the subject's left hand in cold water, her right hand in warm water. After a few minutes, put both hands in a vat of tepid water. The water in the vat will feel warm to the left hand and cool to the right hand.

Such discrepancies afford evidence that there is a subjective element in phenomenal assessments of warmth. Still, the regularities that people discern—such as those that concern water's changes of state and those that correlate with the sequence of the seasons—seem independent of anyone's discerning them. If the subjective factor could be bracketed, the remainder presumably would be whatever mind-independent property underwrites those regularities.

How were they to do it? How, Chang asks, did people invent and calibrate a thermometer when they did not already have one? How could they test it? They needed a benchmark—something to test their device against. That's precisely what they didn't have.

Since subjective assessments of warmth are inconstant and mutually inconsistent, they are not reliable indications of a stable, mind-independent property. They may be true as expressions of the subject's phenomenal state. George's sincere avowal that the wind feels cool to him is true of him. Fred's sincere avowal that it feels warm to him is true of him. But because the claims are at odds, we cannot directly read anything about the wind's objective properties off of them. The avowals cannot both be accurate descriptions of the property of the wind that grounds the two responses. Construed as reports about the wind, the claims conflict.

Rather than jettisoning phenomenal reports of warmth on the basis of such conflicts—declaring them one and all unreliable—we might notice that although some are at odds, a considerable number are in accord. Moreover, there seem to be identifiable circumstances in which many informants

Epistemic Dynamics

give mutually consistent verdicts. And there seem to be identifiable individuals whose judgments regularly and sharply diverge from those of their peers. There are then discernible patterns of agreement and disagreement. These patterns supply evidence.

Suppose we bracket verdicts on which there is no consensus and disregard reports of informants who are frequent outliers. Then let us consider the verdicts that remain. Call these the "clear cases." Many can be put in a linear order. It is widely agreed, let us say, that p is warmer than q, that q is warmer than r, and that p is warmer than r. It is widely agreed that p is not warmer than itself, and that when p is warmer than q, q is not warmer than p. This equips us with the resources to create an ordinal sequence—a scale that enables us to determine that one thing is warmer than, is as warm as, or is less warm than another. Clearly, we cannot locate every assessment of warmth or even every clear case on our scale, but we don't need to. The first step is just to establish a schema that can draw and order crude but consistent distinctions. That sequence affords the conceptual grounding for a thermoscope—a device whose readings correlate with, hence can be seen as measurements of, where things stand on this ordinal scale.

The design of a basic thermoscope is fairly simple. It consists of an ungraduated tube with a liquid in it, where the liquid rises in conditions that people tend to consider warmer and falls in conditions that they tend to consider cooler. The conceptual challenge arises because of our propensity to give inconsistent verdicts. No consistent device can or should track all of the situations where we deem one thing warmer than another. This leads to a difficulty. Given that verdicts about warmth are mutually inconsistent, what warrants the claim that the changes in the height of the liquid correlate with changes in an objective property? The answer is to restrict our attention to the clear cases. If thermoscopic changes correlate with verdicts about clear cases, we have reason to think that the thermoscope tracks whatever the clear cases track. The thermoscope thus is tested against the phenomenal reports on which there is no divergence.

Using a thermoscope, we can reconsider cases we previously set aside. When Fred and George disagreed about whether the tea was warmer than the coffee, both verdicts were set aside. But equipped with the thermoscope, we can test such conflicting judgments. If the column of liquid is higher when plunged into the tea than it is when plunged into the coffee, it vindicates George's judgment. The device, which was initially grounded in phenomenal judgments, therefore has the capacity to adjudicate conflicts in phenomenal judgments. George's opinion aligns with the clear cases. That indicates that he is right. This enables us to augment our set of clear cases and sharpen our distinction. George's judgment can be deemed clearly correct, Fred's, clearly incorrect. The thermoscope thus is not fated to reproduce or ignore antecedent inadequacies. Moreover, the same process provides a way to test thermoscopes against one another. If devices disagree with one another, the ones that match the (current) clear cases will be counted as accurate, the others as inaccurate. The devices now considered accurate will be used to judge the next round of phenomenal reports, yielding a refined collection of clear cases, finding agreements and disagreements that the cruder device could not distinguish. This yields a new class of clear cases, which will serve the standard against which to assess more refined disagreements in thermoscopic readings.

A thermoscope-based account may be in reflective equilibrium. If the devices yield consistent readings that are borne out by their match with clear cases, those readings and the instruments that yielded them are reasonable in light of one another. Ceteris paribus, they can be integrated into a better understanding of warmth than the previous account, which relied entirely on phenomenal reports. They advance understanding by providing verdicts for previously disputed cases. They admit of refinements as the augmented body of clear cases provides data against which attempts to fine-tune thermoscopes can be tested. The process continues, settling the status of increasingly many cases that were in dispute, yielding increasingly refined judgments about whether one thing is warmer than another.

Still, there remain desiderata that thermoscopes cannot satisfy. A thermoscope provides no way to tell how much warmer one thing is than another. Nor does the equilibrium it figures in afford grounds for claiming that changes in the height of the liquid are caused by changes in warmth. So far, we have only a correlation. We want to know how and why the correlation holds. Evidently, an ordinal scale is not enough. We want to know not just that the tea is warmer than the coffee but how much warmer it is. We want a cardinal measure. That requires a thermometer.

To devise one, we need fixed points—ways to anchor the device's scale in conditions that we take to be invariant. After considerable debate and experimentation, scientists settled on the freezing point and the boiling point of water.[5] We also need a benchmark—something to test our new device against. The thermoscope provides one. We engage in the same process used in devising the thermoscope. But rather than testing a candidate thermometer against phenomenal reports, we test it against thermoscopic readings. Having reason to think that the column of liquid rises when conditions are warmer, we can begin to measure (against the two fixed points) how much it rises. We restrict our attention to thermoscopes that agree with one another, setting aside those that give mutually inconsistent verdicts. Once we have what we take to be a trustworthy thermometer, we revisit the inconsistencies we found among thermoscopic readings, just as we used the thermoscope to settle inconsistencies based on phenomenal reports. We vindicate some measuring devices that were originally set aside. We also test thermometers against one another, using the same method. Initially, we will be capable of only relatively crude calibrations. But we can use the process to refine our measuring devices and thereby the measurements we are capable of making. Throughout, there is a dialectical interplay between presumptively plausible readings and presumptively reliable methods, each being subject to revision to accommodate the other. When reflective equilibrium is achieved, the graduated thermometer yields stable, consistent results; it equips us to make objective readings.[6] That objectivity justifies the claim that there is a physical magnitude—temperature—being measured. Changes in the height of the liquid in the thermometer indicate changes in heat.

Not surprisingly—in fact, gratifyingly—further questions arise. How, if at all, are we to project the concept of temperature above and below the fixed points? What justifies the conviction that thermometers that yield consistent readings between 0°C and 100°C also yield consistent readings above or below those points? These are questions that could not have been formulated, much less addressed, until we had thermometers in hand. Once we have a relatively stable concept of temperature and a reasonably reliable way of measuring changes in temperature, we are in a position to ask deeper questions—questions that previously would have made no sense. If temperature is a measure of heat, what is heat? What, that is, is the physical property that we are measuring? Is there also a property of coldness, or is coldness just the absence of heat? At each step, we use the previous stage as a benchmark—that which we test our putative improvements against. We know that the benchmark is to some degree inadequate, but (1) it is the best we have; (2) if we are scrupulous, we can recognize and exploit the information the inadequacies disclose; and (3) over time, epistemic iteration enables us to control for and correct those inadequacies. We accept our results with reservations, being sensitive to exactly where those reservations lie. Through a dialectical interplay that exploits both our agreement and the bases of our disagreement, we stabilize and integrate our commitments, thereby making epistemic progress.

Still, there is a worry. In mathematics, iteration consists in using the output of one step in a process as input into the next step. This might seem to invite the entrenchment of errors. If an untrustworthy mode of inference outputs p, it gives us no reason to accept p. So using that same technique, taking p as an input, should engender no confidence in the product of that inference, q. If we mathematically iterated the results of affirming the consequence, for example, we would just generate further unwarranted conclusions. Mathematical iteration transmits warrant. If the inference pattern is invalid, no warrant is transmitted. Logic and mathematics have the resources for ensuring validity. Elsewhere, the ground is shakier.

This may seem to raise a problem for epistemic iteration. Unlike logicians and mathematicians, investigators working in less regimented fields cannot ensure that the epistemic iterative patterns they generate are valid. Indeed, since we initially accepted some of them only with reservations, we recognize that our inputs and our methods are likely to be inadequate. Epistemic iteration needs to correct inadequacies as well as convey warrant. The epistemic iteration Chang discusses is thus more nuanced than simply taking outputs and using them as inputs. Warrant is augmented through a sequence of carefully choreographed steps. When a disputed opinion is vindicated, it becomes one of the cases against which further refinements will be tested. Thus, as a result of its match with the thermoscopic reading, George's verdict becomes a clear case. Fred's verdict is henceforth deemed clearly wrong. We then use the expanded class to test further refinements of our device. If a refined device accommodates the expanded class, we use it to vindicate some of the cases that remain in dispute. This leads to a further expansion of the class of cases against which we test further refinements. If the device fails to agree with the expanded class, we tinker with it to bring them into accord. Then we test against further disputed cases. There is an oscillating, reciprocal process of assessment, where judgments about individual cases and the emerging technological innovations are tested against one another and each is refined on the basis of what we discover by relying on the other. The process of epistemic iteration thus does more than transmit warrant—it amplifies warrant.

Epistemic Dynamics

It might seem that the process is circular. We test a device under development against clear cases and test cases by seeing whether they accord with the device's output. This is not quite right. Although the cases currently deemed clear and the innovations in instrumentation are tested against each other, both the class of cases and the class of devices (or their calibration) change from one iteration to the next. There is thus no guarantee that any particular attempt at epistemic iteration will succeed. It could turn out, for example, that a class of clear cases resists augmentation. If so, then by incorporating a new case that a device vindicates, we introduce an inconsistency in the class. We may see no way to expand the class of clear cases without introducing such an inconsistency. When this happens, the attempt to use the device to further improve our understanding of the phenomena is blocked. Alternatively, it could turn out that the device does not give consistent readings. In some cases, it deems p warmer than q; in others, q warmer than p. We need not and, at least in the early stages, should not demand complete consistency. We use inconsistency as a way to tell that cases are not clear. But if the vacillation is too great or happens too frequently, there is a problem. Maybe the instrument is unduly sensitive to an unrecognized parameter. Maybe we are taking measurements incorrectly. Maybe despite what we originally thought, there is no objective property to be measured. Epistemic iteration is, as it ought to be, fallible.

My characterization of epistemic iteration is schematic and describes only one simple type. Epistemic iteration need not involve instrumentation. All that is needed is that there be an independent (tentative) criterion of acceptability that we test against clear cases, use to adjudicate disputed cases, then refine to reflect the expanded set of agreed upon verdicts. Nor need there be logical inconsistencies to set the stage. Non-cotenability, inconstancy, or incoherence can play the same role. What matters is that there is a range of agreement and a penumbra of disagreement about a topic and a range of agreement and a penumbra of disagreement in the methods for evaluating judgments about the topic. With the growth of understanding, the iterative process typically becomes more refined. Rather than building exclusively on verdicts on which there is a consensus, it is sometimes preferable to begin with verdicts that strike us as close enough in relevant respects. That would yield a broader and more variegated base. Rather than assigning all lower-order information equal weight, we might develop reasons to think that some bits are more likely to be correct than others. Perhaps we have reason to think that alcohol thermometers are more accurate in a given range than mercury thermometers. If so, we might weigh agreement among alcohol thermometers more heavily, downplaying small discrepancies between them and mercury thermometers. Once we have a viable theory of the phenomenon being measured or the conditions under which measurement is carried out, that theory can put pressure on our instrumentation as well. With the growth of understanding of a range of phenomena comes a growth of understanding of how to understand that range of phenomena. Methods, standards, and criteria emerge in concert with more refined and stable judgments of fact.

Although my discussion focuses on the development of a measuring instrument, it is also a history of the emerging concept of temperature—the magnitude that the thermometer measures—and of the emerging concept and theory of heat—the property that temperature is a magnitude of. Understanding of all three evolved in tandem. Chang's book is properly entitled *Inventing Temperature*, not *Inventing the Thermometer*. As we develop devices to measure things, we learn about the things we measure. And as we learn more about the things we measure, we are in a position to improve our measuring devices. An epistemically fruitful device not only enables us to systematize our understanding it puts us in a position to extend our understanding. We learn something about both water and temperature when we recognize the phenomenon of supercooling—the situation in which water remains liquid below the normal freezing point of water. Without the concept of temperature and the recognition that water freezes at a particular temperature, we would have no reason to think of supercooled water as anything other than very cold water. It would not strike us as exceptional or in need of an explanation. We gain access to conditions on crystallization and insights into the nature of matter when we recognize that supercooled liquids are not just very cold liquids. Here again, epistemic progress gives rise to new questions.

Re-engineering

Epistemic leveraging is not restricted to the developments in instrumentation. Georg Brun (2020) maintains that the evolution of the scientific concept of *piscis*, as a biological kind, starting from the ordinary concept of a fish (which includes all aquatic animals) follows a similar trajectory. He labels this type of epistemic iteration "conceptual re-engineering." The goal is to re-engineer an everyday concept or a concept drawn from a different discipline, by pruning, augmenting, and refining it to render it suitable for a particular scientific purpose. We need not think that the everyday concept is inadequate for its mundane purposes. But it is, we suspect, inadequate for the scientific purposes we want to use it for. Still, it is as yet our best way of marking out the class of entities we want to focus on. We take the everyday concept of fish and exclude animals that strike us as clearly not members of the class we seek to mark out. Thus aquatic mammals, such as walruses and whales, are ruled out. They are clear foils—organisms that we want to bar. We also bracket other aquatic air breathers, not being sure whether they should belong. We suspend judgment about their classification. We then see whether the organisms that remain exhibit enough uniformity to be plausible members of a taxon. We identify properties in the clear cases that seem to qualify them for membership in a single taxon, given what we want the taxon for. Then we go back to see which of the disputed cases share the qualifying properties. The process is iterative. Taxonomists devise criteria for membership that includes all the organisms that are currently uncontroversially considered fish. They then see what else those criteria would apply to. If, for example, the criteria would count lampreys or hagfish as fish and that seems too much of a stretch, they go back and refine their criteria. The goal is not to perfectly fit some antecedent, well-demarcated stereotype but to devise a category that suits a specific range of ichthyological purposes.

Brun uses *piscis* to illustrate the process of conceptual re-engineering. He does not elaborate on the fate of that particular re-engineered concept. It turns out that *piscis* is no longer considered a useful taxonomic category. Ichthyologists can, as Brun says, re-engineer the concept to promote their epistemic ends—but only up to a point, for the world pushes back. If they want a cladistic category—one suited to charting lines of descent—they face the problem that any clade that contains all of what we consider clear cases of fish also contains tetrapods—four-legged mammals and amphibians. That's too comprehensive for their purposes. They are, of course, free to ignore lines of descent and demarcate a kind that rules out tetrapods—that is, to forgo a cladistic classification. There is, after all, reason to think that fish have something biologically important in common with one another that tetrapods do not share. Whether this is reasonable depends on what they want their taxon for.

Still there are difficulties. Even with the exclusion of tetrapods, the class of species that qualify as fish is vast and heterogeneous. Moreover, some species that qualify as fish under the resulting taxonomy are, in biologically significant respects, more like species that do not than they are like other species that qualify as fish. The liberties that underwrite re-engineering enable ichthyologists to draw a line. But they do not ensure that any such line will serve the epistemic purposes that motivate it. The field may reach a point where no foreseeable revision in the concept of fish will promote the relevant epistemic ends. In that case, the next step in epistemic iteration may be to retire the concept—treat not just *fish* but also *piscis* as a "folk concept" without scientific significance.

Retiring the concept would not show that the effort to re-engineer it was epistemically fruitless. Someone like Sellars (1968) might take this view, maintaining that the early iterative steps were mere promissory notes that will be repaid if and only if the successor concept or its descendant proves integral to the understanding that emerges at the end of inquiry. A path that is abandoned then would show itself retrospectively to be an epistemically poor investment. I disagree. Each successful iterative step constituted an advance in the understanding of aquatic animals over that which its predecessors could provide. That, under a given conceptualization, these animals but not those qualify as fish enabled scientists to focus attention on differences in physiology, anatomy, and morphology that were not previously salient. Each step motivates and equips epistemic agents to ask questions that could not previously be asked. Rather than seeing earlier iterations as mere stepping stones to something that will (we hope) eventually amount to understanding, we should recognize that each successful step is itself an advance in understanding. Chang makes a similar point about efforts to devise a perpetual motion machine. They failed. But science learned a lot by recognizing why they failed. The attempts provided the "theoretical foundations of the new science of thermodynamics, in the form of the first and second laws of thermodynamics" (Chang 2022:246).

The possibility of advancing understanding through epistemic iteration does not vindicate all commitments or even all commitments that admit of iteration. Astrology invokes the influence of celestial objects to explain relatively fine-grained events in people's social and emotional lives. The failure of its predictions combined with the vagueness and vacuity of its explanations supply sufficient reason to dismiss the enterprise as epistemically bankrupt.[7] There is no evidence for its claims or for the causal mechanisms that it takes to ground them. Nor would the situation automatically improve if astrological commitments underwent iteration. The mere use of sequences of astrological outputs as astrological inputs would not be enough. What would have to be shown is that the iterations advanced understanding—that the revised concept of astrological influence or of improved methods of discerning such influence yielded better predictions and explanations than its predecessor as well as better predictions and explanations than would have been possible without appeal to celestial influence. Lacking evidence that this is so, there is no ground for thinking that iterations in astrology would be epistemic advances.

In epistemically serious endeavors, such as the development of standards and methods, we can see the trajectory that Chang and Brun describe. Investigators ground their inquiry in what they antecedently take themselves to understand about the phenomena and the available methods for investigating it and assessing their findings. They recognize that the ground is shaky. They realize that current understanding is inadequate for their cognitive purposes. But they think (or anyway hope) it provides a stable enough foothold that they can build on it. The underlying idea is that a network of commitments that is recognized to be flawed can nevertheless provide a suitable platform for improvement. That idea is vindicated when the iterations sharpen distinctions, refine measurements, disaggregate complexes, and provide resources and motivations to ask new questions.

Conditions on Iteration

In Chang's account, the iterative process begins with phenomenal reports of felt warmth. We take it that these are sincere and that the informants arrived at them independently. We also take it that our informants are qualified to make the reports. They have, that is, the requisite conceptual and linguistic resources. Since virtually everyone satisfies these requirements, that the informants are competent seems to go without saying. Assuming these requirements are satisfied, it is reasonable to suppose that the clear cases (at least roughly) track something objective. If, however, the reports were intentionally biased, coerced, or disingenuous, we would have no such reason. The clear cases then would not reveal anything that could plausibly serve as evidence of an objective property. So even at the most basic, pre-theoretical level, informants need to be epistemically responsible. That in turn requires that they be epistemically autonomous.

The iterative process need not start with anything so fundamental as phenomenal reports. Calibration of thermometers was based on thermoscopic readings. We can start with whatever database we consider sufficiently stable. But wherever we start, the data should be, so far as we can tell, untinged by bias, disingenuousness, undue influence, or carelessness. It should be conscientiously generated and reported. Informants then must be epistemically responsible. In phenomenally grounded cases, this just means knowing and accurately reporting on one's own mind. As things get more complex, expertise is needed. Thermometer designers needed to calibrate their devices against the reports of competent, conscientious, sincere thermoscope readers. Even though the data are noisy and some of them will be set aside, because they were responsibly generated and reported, they are a reflection of the current state of understanding of the phenomena under investigation. Such data are plausibly regarded as a sound basis for epistemic iteration.

The process does not presuppose that the clear cases, whether based on popular or expert consensus, are correct. They have a measure of epistemic inertia because they seem the best way to deal with noisy data. Nevertheless, they may turn out to be inaccurate. As we iterate, we refine our resources. Looking back, with our improved understanding, we may see that the original data were not just noisy; they were systematically flawed. The initial phenomenal clear case consensus may have been that Thursday was warmer than Wednesday. But, as it turns out, Thursday was also considerably more humid. Once we have a thermometer, we can discern that Wednesday was in fact warmer, and we can discover that human phenomenal reports are actually reports of a complex of heat and humidity. This is why a 90°F day in Atlanta feels more oppressive than a 90°F day in Phoenix.

When Fred and George made their original assessments, we took it that they disagreed about the wind. That assumption itself may have been unfounded. Perhaps they used different subjective scales. If so, there is a range of felt qualities that Fred calls warm and George calls cool. This possibility is consonant with their both being neurologically normal users of a single language. We need not invoke an analog of an inverted spectrum case to account for the divergence. All that is required is that they differ over the cutoff point. The range of cases that Fred calls warm intersects with but is not identical to the range that George calls warm. If we have no more to go on than their phenomenal reports, we cannot tell whether they disagree about the wind. Similarly, if Alice deems a lamprey a fish and Betty demurs, they might differ

Epistemic Dynamics

over the biological traits of lampreys—a morphological question—or over where the line between fish and non-fish is drawn—a conceptual question. The epistemic advances that Chang and Brun describe enable us to sharpen and refine our distinctions and, over time, replace subjective, qualitative characterizations with more objective, perhaps quantitative, ones. Minimally, this puts us in a position to tell whether and if so about what we disagree. To answer such questions requires a common, intersubjectively assessable metric (see Porter 1996). Through epistemic iteration, we can devise one.

Everyday Leveraging

Epistemic leveraging is not restricted to methodologically sophisticated, disciplinary investigations. It is a familiar strategy in ordinary life. Consider the evolution of a recipe for vegetable soup. The first attempt might consist in simply tossing available vegetables into a pot, covering them with water, perhaps adding a few spices, and cooking for a while. The result is not unpalatable, but it is less than a culinary success. The peas are too mushy, the carrots too hard. There's too much pepper, not enough salt. On the other hand, the potatoes and the broth are delicious. Our next venture makes adjustments to accommodate what we have learned. We don't add peas until the carrots have been cooked for a while. We use a lighter hand on the pepper. We omit oregano entirely. The result, we think, is an improvement, but it could still do with more work. Next time around, we make further modifications—maybe more mushrooms, fewer onions. So far, our efforts involve no measurement. They are qualitative, impressionistic, and vague. In assessing each iteration, we learn to focus on nuances of flavor and texture. Over time we move from tossing ingredients into the pot, adding a dash of this and a pinch of that, to measuring proportions and specifying cooking times and temperatures. We may eventually control for the size or freshness or source of the ingredients. Finally we create something we would be happy simply to repeat. Moreover, the resulting recipe is one we would be able and willing to transmit to other cooks. It has sufficient precision that others could replicate our soup. Further iterations may be in the offing, but for now we have a recipe we are satisfied with.

Epistemic iteration involves feedback loops. It might seem that they are absent in this case. But the process involves more than merely changing ingredients and cooking times. While improving the soup, we also refine our palate, developing a capacity to discern and evaluate nuances of flavor that we were previously insensitive to. We refine the soup and our standards for a palatable soup in tandem. Recipe, discernment, and standards evolve together, each adjusted in response to the others.

Still, one might wonder what this has to do with epistemology. Developing a recipe is surely a practical matter with culinary, not epistemic objectives. But even though the dominant objective is culinary, there is an important epistemic component. The rationale for the iterative process is twofold. We want a better soup next time around. We also want to know how to consistently make a tasty soup. A one-off success is not good enough, for our goal is not just a tasty soup today but a procedure that we can rely on to produce tasty soups in the future. That's an inductive issue. It involves drawing on background knowledge, developing new knowledge about ingredients and cooking times, and integrating these considerations into a network of culinary epistemic commitments. It involves figuring out which variations matter and which ones don't. It involves discovering permissible substitutions and their effects. Although many of our everyday activities are not purely epistemic, they have irreducibly epistemic elements.

Making Progress

Progress rarely happens by accident. Granted, things sometimes fortuitously turn out well. But normally, if there is progress, it is because we *make* progress. We intentionally facilitate improvement. One important way that we make epistemic progress is by structuring our inquiries and designing their institutional supports in such a way that current understanding can be leveraged to yield further understanding. This involves epistemic iteration. Thus, as we have seen, an understanding of temperature can start from nothing more than felt differences between warm and cool systematize, refine, and revise them to ensure consistency, and end up with a thermometer capable of correcting the judgments we make on the basis of phenomenal differences in warmth, and with a magnitude—temperature—which is independent of subjective opinions.

That epistemic progress takes this form is not inevitable. We could structure inquiry in such a way that whenever we reach an impasse, we simply throw everything out and start again from scratch. In that case, all that we

Epistemic Dynamics

would glean from a failed inquiry is that a given approach faced seemingly insuperable barriers or that it did not go far enough. We would not build on or improve the understanding we had; we would simply jettison it and try something else. Except perhaps when we're hopelessly flailing about, this is neither what we do nor what we should do. What we do, and should do, is exploit our epistemic successes and failures, designing our inquiries and networks of commitments to enable us to do so.

The issue is indifferent to the realism/pragmatism/antirealism debate. A realist thinks that the overarching goal of inquiry is to discover the true and ultimate structure of reality. A Deweyan pragmatist thinks that inquiry is endless, that every level of epistemic achievement sets the stage for further inquiry. A constructive empiricist thinks that the goal is empirical adequacy. All three face the question of how to improve on current understanding in order to achieve their goal.

Understanding comes in degrees. One understanding of a topic can be better than another. The expert, we think, understands her topic better than the novice does. Later science, we think, affords a better understanding of the phenomena it investigates than its predecessors did. To make sense of this, it would help to have a criterion for what it is that makes one understanding of a topic *better* than another.

Elsewhere, I argue (Elgin 2017:57–62) that objectual understanding is non-factive. So I cannot endorse a criterion like:

Theory A affords a better understanding of the phenomena than *Theory B* because *Theory A* is committed to more truths than *Theory B*.

Factivists ought not endorse it either. If closure obtains, any theory that is committed to a single truth is committed to infinitely many truths, for from one truth, infinitely many others logically follow. So assuming they both contain at least one truth, *Theory A* and *Theory B* are committed to exactly the same number of truths. Even if closure does not obtain, there are apt to be difficulties. Kuhn contends that the Ptolemaic alternative was as accurate as Copernicus's theory (1970:75–76). Arguably, that means that the two were committed to equally many truths. But we would not want to say that the two provided equally good understandings of celestial motion. In any case, counting truths is not a promising strategy (see Treanor 2013). Even if we could figure out how many truths a given theory is committed to, we would still want to assess their significance. If *Theory A* is shallow and *Theory B* is deep then, other things equal, there are apt to be good grounds for thinking that B affords a better understanding than A does, even if A embodies a larger number of individual (albeit rather superficial) truths.

Some philosophers maintain that understanding is tied to explanation. To understand a topic is to be able to explain why the phenomena behave as they do (see Grimm 2006; Hills 2016; Khalifa 2017). As I argued earlier, I disagree. Like Lipton (2009), I think it is possible for a person to understand a phenomenon and to be able to display her understanding, even though she cannot explain the phenomenon. In any case, explanationist accounts of understanding do no better than non-explanationist accounts at showing what makes one understanding of a range of phenomena better than another, assuming both afford *some* explanation of the phenomena they pertain to. Criteria for determining what makes one explanation better than another face exactly the same problems as criteria for assessing the merits of different understandings. Kuhn (1970) maintains Ptolemaic astronomy provided as good (or as bad) explanations of the motions of celestial objects as the original Copernican theory did. Nor can we break the tie by insisting that the Copernican explanation was true, while the Ptolemaic one was false. Neither is true. Nevertheless, we rightly want to insist that the Copernican theory affords a better understanding.

To be sure, even without a criterion, we have bases for comparison. Depth, as I said, is one. An expert's understanding of a topic is deeper than a novice's, in that the expert appreciates underlying grounds. Although the novice grasps connections between phenomena, she has little appreciation of what grounds them. In respect of depth, the expert's understanding is better.

Breadth is a second factor. One person's understanding of a topic is broader than another's if it comprehends a wider range of phenomena. Thus, for example, an ornithologist who understands the migratory behavior of birds in general has a broader understanding than a specialist who understands only the migratory behavior of Canadian geese. In respect of breadth, the generalist's understanding is better than the specialist's.

Significance is a third. Even if an expert and a novice accept the same account, they may differ in the significance they attach to the information it contains. Both an expert and a novice historian might recognize that in the Battle of Marathon, the Athenian soldiers were uncharacteristically arrayed so that the battle line had a weak center and strong flanks. The novice makes nothing of this fact, maybe thinking that the soldiers had to line up

Epistemic Dynamics 111

somehow, maybe not thinking about it at all. The expert recognizes that the configuration of the line was crucial to the Athenian victory, as the strong flanks enabled the Athenians to surround the Persian Army when the weak center retreated.

A fourth factor is systematicity. Understanding involves a network of cognitive commitments. Ceteris paribus, an account that tightly connects two bodies of information via multiple strands affords a better understanding of the topic as a whole than one where there are few and/or weak links between them. A comprehensive, integrated history of the migration from Europe to North America is in respect of systematicity better than one that simply tacks an account of the migration from Southern Europe onto an independent account of the migration from Northern Europe but uses nothing stronger than classical conjunction to link the two (see Dellsén 2021).

Moreover, *Theory A's* having fewer significant errors than *Theory B* contributes to *Theory A's* claim to afford a better understanding of their shared subject matter. *Theory A's* having fewer gaps than *Theory B* would also give *Theory A* an edge. Perhaps if *Theory A* is more elegant than *Theory B*—if it contains fewer idle wheels—that too would be a reason to think that *Theory A's* understanding is better. The list goes on.

Better in a single respect is not, of course, better overall. One difficulty in making comparisons is that there are trade-offs among the various factors. If one theory affords a deeper understanding and another a broader one, it is not obvious whether or why one is better than the other. If the overtly fictive elements in *Theory B* enable it to comprehend a greater range of cases or to display an abstract structure that the phenomena share with another seemingly remote range of phenomena, then the fictive elements seem to be an asset rather than a liability. They might, for example, disclose structural parallels between two apparently distinct ranges of phenomena. The idle wheels in Maxwell's model enable it to display an analogy between electromagnetic and general dynamic processes. The model thereby shows a uniformity at a higher level of abstraction. This is a signal achievement (see Nersessian 2008:19–60).

For our purposes, a further difficulty is that the foregoing features just characterize the grounds for preferring one theory to another. The two accounts—for example, two histories of the Battle of Trafalgar—could be completely independent of one another. One might focus on naval architecture—the designs of the ships of the two navies; the other, on military strategy—the

112   Chapter 4

battle plans of the two forces. The features I've listed indicate nothing about what it takes for an understanding to *improve*—about what it takes to make epistemic progress. Because understanding is always limited and usually less than ideal, a critical question is whether a network of commitments that embodies the current understanding of a range of phenomena supplies resources for improvement.

The discovery of the structure of DNA did not just supply an answer to an outstanding question; it did not just disclose a previously unknown fact. It provided a springboard for far-reaching advances in biochemistry. It effectively gave birth to the science of molecular genetics. Such findings are often said to be fruitful. But a finding, however impressive, is not fruitful if it falls on barren ground. To bear fruit, it must be incorporated into a nurturing epistemic environment—a network of commitments that can foster its elaboration and extrapolation in a way that leverages the finding to promote or reorient inquiry. One reason Watson and Crick's description of DNA's structure was fruitful was that mid-twentieth-century genetics was poised to receive it.

Standardly we tend to think that understanding progresses when problems are solved and questions answered. This is so, but I want to suggest that the dynamic is more complicated. Findings are fruitful not only because they answer questions that we previously could not answer or solve problems that we previously could not solve. They also equip us to *ask* questions that we could not previously ask and *pose* problems that we could not previously pose. Watson and Crick's finding equipped geneticists with the ability to seriously investigate, rather than merely wonder, how genetic information gets transferred from one generation to the next and to seriously investigate, rather than merely wonder, how mutations happen. The mark of a good answer is that it raises good questions (see Dewey 1916:339–341). I suggest that for a finding to be fruitful, it has to be more than just valuable in itself. It has to fit reasonably well with or successfully challenge what is currently understood. Either way, it has to point beyond current understanding. It must provide an avenue for epistemic progress.

Humans are an inquisitive lot; we are rarely satisfied. We have a variety of cognitive itches we'd like to scratch and cognitive knots we'd like to untangle. To keep terminology simple, I'll call all of these "quandaries." Let's sort them by degree of difficulty.

Some are relatively easy. We have what we need to answer them. I do not know how to drive from Atlanta to Birmingham, but it wouldn't be hard to

Epistemic Dynamics

113

find out. I need only look at a suitable map. Nor do I know the square root of seventeen, but I would have little difficulty calculating it. Let us call such straightforwardly settled quandaries "puzzles."

Not all puzzles are grounded in a particular individual's epistemic shortfalls. There are plenty of puzzles that no one has solved. Perhaps no one has calculated the cube root of 378,953,929. Still, it's not that hard. Someone surely could do the calculation. If it's never been done, that's just because no one had the incentive to make the effort. Perhaps no one has ever tried to figure out how to drive from Atlanta to Birmingham via Duluth (which is a couple of thousand miles out of the way). But it would be straightforward to do so. A quandary qualifies as a puzzle on this usage just in case the resources required to resolve it are at hand or are readily accessible.

Puzzles need not be trivial. Kuhn (1970) maintains that the business of normal science is puzzle solving. He does not think that normal scientific puzzles are all easy. His point is only that the methods, techniques, and resources for solving them are taken to be available. If Kuhn is right, scientists doing normal science assume that current science supplies what they need to solve the puzzles they confront.

At the opposite extreme are what I will call "conundrums." They are questions we might like to answer, but we take it that we do not now have, and have little if any idea how to get, the resources needed to answer them. Maybe we would like to understand the nature of consciousness or to know what existed before the Big Bang or to figure out how to enable people to travel from one galaxy to another. But right now, such matters are apparently beyond our epistemic reach.

Not all conundrums are gaping holes in the fabric of understanding. In some cases we have some (or maybe even all) of the resources we think we need but do not know how to exploit them. I suspect that mathematics has the resources to settle the Goldbach conjecture. Still, no one knows how to do it. All the obvious, seemingly plausible approaches have failed. This was the situation regarding Fermat's Last Theorem until Andrew Wiles creatively wove together a wide array of antecedently available mathematical threads to provide a proof. The problem was so difficult, not because math lacked the requisite resources but because there was a vast gulf between the discipline's having the resources and anyone's having a clue how to exploit them. We need not assume that conundrums are in principle unanswerable, but we construe a quandary as a conundrum when we suspect that it is currently intractable, and we treat it as such.

In between are the predicaments I will call "problems." These are quandaries we do not currently have (or do not think we currently have) the epistemic resources to answer. But we have (and at least suspect that we have) the epistemic resources to develop the resources to answer them. This may involve new techniques, new orientations, new devices, new measurement procedures. One perennial epistemic objective is to convert conundrums into problems and problems into puzzles. Then, of course, we have to solve them.

As I've described things, inquiry may sound like an extended exam. The questions are set out before us. We need to answer them—if we can—exactly as they are formulated. Question 1 is a puzzle; it shouldn't be too hard. Question 2 is a problem; it will take more time and perhaps a bit of innovation. Question 3 is a conundrum; don't even bother trying to solve it unless you have time to spare. This is not quite right. One reason is that my taxonomy is vague. Which side of the line a particular question falls on is not always clear. Some philosophers of mind think that uncovering the nature of consciousness is a problem, not a conundrum. Although they grant that it is difficult and the route to a solution is far from obvious, not only do they think that it can be solved, they also think that they have or can devise the resources to solve it.

The main objection to the exam analogy has nothing to do with vagueness though. It is that our epistemic predicament is not a matter of nature *confronting* us with puzzles, problems, or conundrums that we are then expected to deal with. We *contrive* them. So rather than asking whether a particular quandary *is* a puzzle or a problem or a conundrum, we should focus on what it is to *treat it as* one sort of quandary or another. To treat something as a puzzle is to treat it as something that we pretty much know how to solve as it stands. To treat it as a problem is to treat it as something where the available resources might need to be extended or modified, or the problem itself might need to be tweaked to yield an acceptable solution. To treat it as a conundrum is to admit that it is, as far as we can tell, currently beyond us. Nor do we just *treat* a particular quandary as one sort of quandary or another. We *frame* it in a particular way, so that it admits of being treated as a particular sort of quandary.

We have a variety of strategies for doing this. Sometimes framing an issue as a problem instead of a conundrum is a matter of construing it in terms

Epistemic Dynamics

of relevant alternatives rather than open-endedly. A quandary is more tractable if it is a matter of asking whether x, y, or z accounts for φ rather than asking what in the entire cosmos accounts for φ. Sometimes, framing it as a puzzle rather than a problem is a matter of setting lower bounds on precision. Although we can't say exactly how fast a particular virus replicates, we can determine that its replication rate falls within a specified interval. Sometimes, framing it as a conundrum is a matter of requiring a more exact solution than we are in a position to provide—for example, demanding an analytical solution rather than accepting a statistical one. Correlation does not imply causation, we insist; but we recognize that we're clueless about how to find the missing causal link. By setting the conditions on what will count as an acceptable answer, we fix the character of the quandary.

Linus Pauling thought that the structure of DNA was a single helix. Had he been right, the structure of the molecule would not, or at least not so directly, have suggested a route to understanding genetic transmission. It would not have supplied as many or as clear clues as the double helix does about the way genetic recombination occurs. The structure of DNA is whatever it is. Epistemologists cannot wander in and announce that we'd prefer it to have a more likable structure. But the issue here is not what the fact is. It is what to *make of* the fact. The fruitfulness of a finding depends on what we can make of it. And what we can make of it depends on the dynamics of the understanding that will incorporate it.

To understand a topic is to reflectively endorse a network of cognitive commitments in reflective equilibrium. Sometimes we are satisfied with a static equilibrium. We have a difficulty that needs to be addressed, but we do not expect it to recur. Suppose Ben wants to know how to make scrambled eggs on a camp stove. The instructions the cookbook provides are clear and complete. The understanding that emerges is a network of culinary commitments in reflective equilibrium. Conditions are sufficiently stable that there is no realistic worry about the equilibrium becoming unsettled any time soon; nor is there a worry that their adequacy being responsibly called into question on reflection. Since he is not planning to make a career of cooking eggs outdoors, he need not worry about innovations in stoves or evolutionary pressures on eggs. A static equilibrium is all that is needed.

Here, however, we are concerned with how an understanding evolves. Even the most stable systematic understanding is apt to be buffeted by challenges. Further evidence, new methods, considerations drawn from adjacent

116   Chapter 4

fields, novel perspectives, even off-the-wall objections may call accepted commitments into question. What should we make of them? What stance should we take toward them?

One option is to adopt a strong defensive position. In that case we design our networks to absorb what can be readily accommodated and reject everything else as mistaken, misguided, or misleading. This is the stance of the dogmatist Kripke (2011) describes.[8] Since she knows that p, p is true. Since p is true, any evidence against p is misleading. Since she does not want to be duped into losing knowledge by accepting misleading evidence, she therefore rejects any evidence against p. She closes herself off from challenges to p. Even if this were to be a viable stance for someone who knew and knew that she knew that p, for the issue that confronts us here, such a stance is not viable.

One reason is that the criteria for reflective equilibrium do not come close to showing that a network in reflective equilibrium satisfies the conditions on knowledge. Our evidence may be spotty or sparse. Our methods are far from fail-safe. Our modes of inference may be flawed or less powerful than we wish. Many of the truth-apt elements in our theory are not, and are not supposed to be, true. What is to be said in favor of these commitments is that, taken collectively, we have no better. Reflective equilibrium assures us that the considerations are reasonable in the epistemic circumstances. But real-world epistemic circumstances are less than ideal.

Another reason is the pessimistic meta-induction. A quick glance at the history of thought reveals that networks of commitments that once satisfied the criteria for reflective equilibrium have been justifiably overturned. What was reasonable in one set of epistemic circumstances turned out not to be reasonable when circumstances changed—when instruments improved, when more data were gathered, when inferences were refined, when criteria for evidence were sharpened, when cogent unforeseen objections were raised, when our purview is expanded. It is intellectual hubris to think that we are less vulnerable than our forebears.

The Kripkean dogmatist's stance is doctrinaire. It ensures the stability of current understanding by rejecting anything that fails to fit with what is already accepted or by warping new inputs so as to force them to fit. New inputs must align with the received doctrine rather than the doctrine and the new inputs being adjusted to fit one another. Those who take such a stance are so convinced of the adequacy of their current understanding that they

Epistemic Dynamics

117

endorse criteria of acceptance that stifle innovation. This, I suggest, is the stance of fundamentalists. Fundamentalism, as I use the term, is not exclusively a religious stance. What is critical is that core commitments are held to be invulnerable to criticism, correction, or rejection. All revision is to be deflected to other channels. Fundamentalist core commitments then have maximal epistemic inertia. They are immovable. Such a posture does not preclude progress. But in mandating that certain commitments are irrevocable, it insulates them from investigation and assessment. In taking them to be irrevocably settled, it closes off avenues of investigation. In so doing, it impedes progress and preempts the possibility of discovering that the core commitments are wrong. I will have more to say about the perils of preemption in chapter 7.

A more judicious stance acknowledges our limitations. It recognizes that the best we can do in the current epistemic circumstances is not necessarily the best we can ever do. It grants that what satisfies current standards might nevertheless be wrong. And it recognizes that current understanding, however well it stands up to testing, is not the whole story. Moreover, it takes both the fallibility and incompleteness of current understanding to be epistemic assets rather than liabilities.

When we settle an issue, we consider it closed. We see no need to investigate it any further. But if we consider the door still open, even if only slightly ajar, we are receptive to further inputs. Nor should we be grudgingly receptive. We take it to be our responsibility to seek out potentially worrisome inputs. We build corrigibility and expandability into our networks of cognitive commitments. We provide incentives to identify vulnerabilities and opportunities, and we integrate resources for accommodating them into our networks. If it were wrong, where would it go wrong? How would we tell whether it had gone wrong? If it could be expanded, what would we need to prompt a viable expansion? How would we recognize such a prompt? How might we exploit it?

The difference between the stances holds whether or not there is any actual inadequacy in the current understanding of a topic. Even if our current understanding is a stable equilibrium that affords a comprehensive grasp of its subject matter, the two stances diverge. The judicious stance, on the lookout for obstacles and opportunities, entertains and accommodates a host of counterfactuals. It continually asks "What if?" It locates the phenomena and the understanding of them in a richer logical and conceptual space than the doctrinaire stance does. As a result, it is apt to yield a far greater appreciation of the weave of the current network.

A worry arises. The doctrinaire stance is dogmatic, self-protective, and resistant to innovation. It seems reluctant to supply vehicles for epistemic progress. But the so-called judicious stance may appear excessively open-minded. What ensures that it will not be vulnerable to endorsing cavalier revisions—to cheerfully adapting itself to accommodate any passing intellectual fad? If everything is, in principle, open to revision, how can it police candidates for revision? How can we avoid epistemic gullibility? I suggest that we already build mechanisms for impeding gullibility into our epistemic practices.

Familiar methodological protections provide constraints against cavalier alterations. Results, we insist, must be reproducible and should be confirmed via a variety of tests or observed from multiple points of view. Findings should hold up under peer review. Accounts in adjacent domains should mesh. And so forth. Our reasons for granting such commitments considerable epistemic inertia lies in their track records. Systems that respect such constraints tend to be both stable and fruitful. Systems that flout them tend not to be. The normal protections against carelessness, bias, lack of rigor, and other forms of intellectual irresponsibility also protect against gullibility. A commitment that satisfies these conditions could still be wrong. The most we can say about a commitment with considerable inertia is that if we are wrong about it, then we're wrong about a lot. That may give us reason to suspect that we are not wrong about it—reason that justifies attempts to preserve it through revisions. But, of course, we could be wrong about it because we could be wrong about a lot. There are no guarantees.

Something of epistemic value is lost when we treat an understanding of a topic as immune to rejection or correction or elaboration. One thing that is at least diminished if not entirely lost is the opportunity to leverage our understanding. If we take it for granted that our position vis-à-vis p is epistemically adequate, we have no incentive to investigate the grounds for p, to entertain alternatives to p, or to query the parameters within which p holds. Indeed, if we really think it is settled, we ought not look further. The conclusion needs no further support. So additional evidence in its favor is redundant and negative evidence is, we are convinced, misleading. If we think it might be inadequate, we have incentives to look further. Even if we conclude that we are not wrong to endorse p, we are apt to understand p and the

Epistemic Dynamics

network it belongs to better because we have investigated the matter more fully. Acceptance should not be dogmatic. Nor should a network be so self-protective that it simply rebuffs challenges. We should be fallibilists not only because we might be wrong but also because fallibilism promotes enriching our understanding, exploiting both our assets and liabilities in order to make further progress. By that, I do not mean that we should reluctantly, hand-wavingly concede the possibility that we're wrong somewhere. We should be cognizant of where and how we might be wrong and of where and how, even if we are not wrong, the position could be extended, deepened, strengthened, or refined. That is, we should be sensitive to the vulnerabilities in and limitations of the accounts we currently accept, and we should integrate into our network both checks on acceptability and ways to recognize errors. We should also integrate into our network ways to amplify, expand, or deepen the understanding it currently provides.

Epistemic progress often comes about by leveraging current understanding. A network of commitments in reflective equilibrium is as reasonable as any available alternative in the epistemic circumstances. This does not make it optimal. Probably it is not. The most we can say for it is that it is as good as any available alternative. So the understanding it affords can in principle be improved upon. Being limited in scope, it should admit of expansion. Being apt to have flaws, it should incorporate mechanisms for their detection and correction. When we seek to leverage current understanding rather than scrap it and start over, our goal should be to, as far as possible, preserve its successes and remedy its shortfalls. Fruitful findings (as well as fruitful methods, devices, orientations, and the like) are epistemically valuable because they not only afford but *highlight* opportunities. "The specific pairing we have postulated *immediately suggests* . . . ," say Watson and Crick (1953:738). That is, the finding itself points the way forward. We don't need Watson and Crick to tell us. The finding itself exemplifies the opportunity. It orients us toward a potentially (and, in this case, actually) viable route to further understanding.

Ideally, a network of commitments should be primed to exploit success. A good shot in billiards not only scores, it sets up the table for subsequent shots. Something similar holds in inquiry. An understanding of a topic makes sense of a range of phenomena. But it should also be configured so as to promote further understanding. The current cluster of commitments should invite elaboration, correction, and expansion so that it fosters further understanding.

Conclusion

The process of achieving reflective equilibrium is a matter of continuing to revise commitments to bring them into accord with one another and to bring the resulting system of commitments into as good an accord with our initial commitments as any available alternative. Inasmuch as we recognize that we are neither infallible nor omniscient, we take it that the revisions we are currently undertaking are unlikely to be the last word. Rather, they should provide a relatively stable platform for further revisions. In deciding what considerations to revise and what revisions to make, we consider both how the system as a whole answers to our antecedent commitments and what we hope to do with our re-systematized understanding. Some of our commitments are forward looking and pragmatic. There is nothing arbitrary about our choice of revisions. A network that is congenial to leveraging is purposefully designed to provide a platform for its own improvement. Epistemic iteration is a self-correcting procedure. It brackets inconsistencies but neither discards nor disregards them. Instead, it pinpoints and exploits error—using it as a resource to extend understanding. A sophisticated scientific system, like the one that led to the emergence of temperature, might include rigorous methods for rooting out infelicities and identifying opportunities. A mundane one, like the one that led to the recipe for vegetable soup, might do little more than register what was done, what resulted, and how satisfactory the results were. Both processes are iterative and progressive. Both enable us to start from what was unsatisfactory and improve our epistemic lot. Neither requires acceptance with reservations, but both are congenial to it.

5 Realms of Epistemic Ends

I argued earlier that epistemic autonomy and interdependence are mutually supportive. If an isolated agent satisfies the standards that she sets for herself, her opinion is subjectively justified. But because an individual's standards may be idiosyncratic, merely subjective justification is weak. When her opinion satisfies community standards, worries about idiosyncrasy wane. How much this enhances her position depends on the nature of the community that backs her. Unless most of its members function autonomously, consensus counts for little. If, for example, the community values accord more than understanding, members may judge considerations on the basis of their popularity rather than their tenability. Mere consensus does little to augment justification.

Intersubjective agreement among autonomous agents pays dividends. Endorsement from multiple perspectives stabilizes an individual's epistemic commitments. A single line of justification is precarious. All is well so long as that line holds, but if it frays—if, for example, particular bits or sources of evidence are undermined—confidence in the commitment it supports should wane. A network whose nodes are sustained by multiple lines of support can withstand such challenges. Consensus of autonomous compatriots, particularly of compatriots who rely on different reasons, lessens vulnerability. The recognition that one's peers would raise objections if they considered a consideration unfounded justifiably strengthens an agent's confidence. Their silence, as well as their express agreement, gives her grounds for thinking that she is on solid ground. Cooperation, moreover, enables amplification of efforts. Collectively, a group can do more than any one member could do alone. A team can, for example, conduct a more extensive survey than any single investigator could. This would supply more evidence, strengthening the support for a claim. This advantage is consonant with the idea that the epistemic benefits of cooperation consist in doing better at performing antecedently definable and individually pursuable cognitive tasks. A team might glean a large body of evidence by doing a wide-ranging observational study that, but for practical limitations, an agent could do on her own. Let us say that such doing more of the same *extends* an agent's epistemic range. This, no doubt, is valuable.

But epistemic communities do more. They design and engage in practices that configure the terrain in ways that make new sorts of epistemic success possible. They not only extend the epistemic range they also *expand* it. They devise and collectively endorse particular methods and standards to serve their epistemic purposes. They develop and implement new conceptual, inferential, material, and/or technical resources. A community expands its resources by validating methods, systematizing measurements, and calibrating instruments. Through a series of iterations, it increases its understanding of how to understand its domain. Members of the community recognize one another as individually and collectively responsible for setting and promoting their epistemic ends; consensus supplies each member with a measure of reason to think that the goal is met.

Nevertheless, the resulting justification may seem disappointingly shaky. The worry is this: although judgments compatriots agree on are not idiosyncratic—although they satisfy standards collectively *designed to* promote certain ends—those standards still might be ill-attuned to the goal. Satisfying the standards they jointly accept could fail to promote the ends they seek to promote. More needs to be said both about the individual and about the community to allay this concern.

Communities

A community is not an unruly mob. It consists of members who are, and take themselves to be, accountable to one another on the basis of shared expectations (see Goldberg 2018). Nations are communities whose citizens are accountable to one another under law; corporations, communities whose shareholders and stakeholders are accountable to one another under the articles of incorporation; armies, communities whose officers and enlisted personnel are accountable to one another under the military code. Social groups such as clubs, teams, orchestras, and neighborhood associations are communities whose rules, whether tacit or explicit, determine what members can

Realms of Epistemic Ends

reasonably expect of one another thus when and how members can fault one another for failures to perform appropriately.[1] Members of both formal and informal communities are accountable to one another under a variety of (often implicit) codes of conduct as well. Communities thus constitute themselves as realms of ends.

The expectations that bind compatriots are normative, not necessarily predictive or descriptive. They set standards for appropriate behavior, and they retain their authority even when the standards are violated. If I tell my students that I expect their papers by Friday, my statement is normatively appropriate, even though empirical evidence strongly suggests that a considerable number of them will fail to meet the deadline. If my expectation purported to predict, the large number of scofflaws would be evidence that it was unwarranted. If it purported to describe my mental state, the large number of scofflaws would be evidence that my view of the situation was misguided. Since it is normative, the expectation retains its authority, putting the scofflaws in the wrong.

The rationale for a community's standards typically depends on the ends it seeks to promote—to turn a profit, enact a law, understand a phenomenon, or just pass an enjoyable evening with friends. (The regulars down at the pub constitute an informal community, establishing local norms for acceptable barroom behavior.) The standards provide a basis for members to assess one another's actions and to sanction those who fail to act appropriately. Even if, from a purely culinary point of view, the potluck dinner would be better without my tofu delight, members of the neighborhood association are entitled to blame me if I violated our joint agreement that everyone would contribute a dish. My omission, although perhaps secretly welcomed by those with refined palates, is culpable given the agreement I was party to.

As I've characterized it, a community is a Kantian realm—"a systematic union of different rational beings through common laws" (1981:§433). It is a not just a *union* of rational agents, it is a *systematic union*. Systematicity provides structure, enabling a community to devise practices that promote its ends.

Practices

A practice is a network of norms, standards, criteria, and rules that defines ends and sets constraints on the means by which those ends can permissibly be realized. By imposing them on its members, a community structures their joint ventures and furthers their collective aims (Elgin 1996:60). An epistemic practice is one that defines and promotes the realization of epistemic ends. By specifying what is permitted, what is obligatory, and what is prohibited, a practice creates and structures a space of alternatives. A logic, for example, is a practice that specifies conditions on validity. Apart from syntactical constraints, it is indifferent to the content of an argument's premises. But it mandates that only sequences that satisfy its rules are valid. It thus sets constraints and creates a space of possibilities for satisfying them.

Practices underwrite incentives and priorities. They make it the case that some choices are better than others. Chess incentivizes protecting one's queen. Wanton indifference to the queen's vulnerability is apt to be catastrophic. Biomedicine may prefer enzyme assays that yield false positives to ones that yield false negatives when wholly accurate measures cannot be achieved. If one wants to know whether a mathematical claim is true, a proof by reductio might be fine; if one wants to know why it is true, a direct proof is preferable. Such incentives and priorities orient practitioners to some aspects of a situation and occlude or marginalize others. By fixing practice-specific ends and means, they constitute distinctive sorts of success and failure. A logic makes it possible both to prove a theorem and to commit a fallacy. Absent a logic, one can do neither.

Characterizing practices as sets of constraints suggests that they limit freedom. To an extent, they do. An agent is not playing chess unless her relevant actions are constrained by the rules of the game. Of course, she is not obliged to play chess. If she wants to move what are conventionally construed as chess pieces around a board in an idiosyncratic fashion, she is free to do so. If she wants to contrive a different game using the same pieces, that option is open as well. Similarly, a reasoner is not making a classical inference unless she respects disjunction introduction. But she is not obliged to make a classical inference. If she wants to engage in unconstrained conjecture, she is free to do so. If she wants to invent or deploy an alternative logic, she can do that too. Just as the rules of chess constrain the player's behavior only insofar as she is playing chess, the rules of classical logic constrain her only insofar as she is doing classical logic.

The constraints that a practice sets are also enablers. Only by abiding by them is it possible to make the choices the practice defines and to achieve the goals it aims at. The rules of chess make it possible to checkmate. There is no way to literally achieve that goal outside of the game. Nor is there any way outside the game to choose to sacrifice one's knight in pursuit of that goal. The rules of logic make it possible to prove a theorem. Without the rules, no sequence of inscriptions would qualify as a proof. Nor would any sequence qualify as a fallacy. Practices create the possibility for achievements and errors that could not otherwise occur.

A practice expressly weaves together a variety of components to frame activity and provide structure. Can a law be enacted in the absence of a quorum? Does p entail $p \lor q$? In the construction of a practice, answers to such questions are initially matters of choice. But the choices are rarely arbitrary. Designers consider what ends they want the practice to promote and what resources are available (see Rawls 1999). Game designers may take the end to be enjoyment, and they may seek, through their design, to achieve a balance of challenges and rewards that participants will find gratifying (see Nguyen 2020:5–11). Designers of a legislative practice face a delicate issue of balancing individual rights with collective goods, melding political ideals with quotidian demands. Designers of an ethnographic practice face the question of how much scope to sacrifice to increase detail.

Although there are multiple dimensions along which choice is possible, not every element of a practice is intentionally chosen. Unintended and unforeseen commitments may be catalyzed by those that were intentionally incorporated. A practice designed to protect human research subjects might mandate that studies require the signed consent of those being studied. If so, it precludes research on people whose language lacks a written alphabet. Whether or not the prohibition was intended, so long as the rule is in effect, the review board has no choice but to reject proposals to study members of such groups. Once a practice is in place, its constraints structure the activities it licenses.

Rawls (1999) distinguishes between the good(s) *of* a practice and the good(s) *in* a practice. The goods *in* a practice are constituted by the practice. The final good in chess is checkmate; a subsidiary good is capturing one's opponent's knight. Outside of the practice, there may be nothing to recommend them. Their status as goods is internal to the game. The good *of* a practice is external. It is an independently characterized goal that the practice as a whole seeks to promote. The good of a game may be the enjoyment of overcoming obstacles one has set for oneself (see Nguyen 2020:11–13); the good of a legislative practice may be a system of just, enforceable laws; the good of an epistemic practice is to afford understanding of a particular domain. Although a practice is designed to promote a particular external good, this does not guarantee that it will succeed. Matrix mechanics was designed to yield an understanding of the quantum realm. But the calculations were so complicated that even top physicists could not do them or grasp their import (see De Regt 2017:246–251). The complex calculations that were goods in the practice turned out not to achieve or promote the good of the practice.

Although I have been speaking as though a practice constitutes a static network of commitments, many practices are dynamic. Among their commitments are rules of revision—meta-principles that dictate how and when to revise their first-order commitments. Arguably, the rules of chess are static or at least evolve very slowly. But the rules of legislative practice are dynamic. They allow for rule-governed revisions in how laws are to be made. Appreciating that we are finite and fallible, epistemic communities favor dynamic practices. Rules of evidence and inference, standards of acceptance, and methods of inquiry are subject to revision. Let's look at a case.

The Royal Society was founded in 1660 with the objective of advancing understanding of the natural world. It constituted itself as a community of natural philosophers—a realm of epistemic ends. The goods in the practice consisted of fruits of the methods, techniques, and norms it devised and endorsed, as well as the opportunities to report one's findings. These were believed to promote the Society's end.

Initially the Royal Society admitted only English gentlemen (see Hunter 1982; Shapin 1994). This was not pure snobbery. The rationale for the restriction was the conviction that a gentleman, being beholden to no one, could be trusted. His word was his bond. Having sufficient wealth, power, and prestige, he would not be swayed by political or pecuniary considerations. "It was the disinterestedness of the English gentleman's situation that was most importantly identified as the basis of his truth-telling. The specific circumstances of his economic, political, and social free action . . . were mobilized as explanations of the integrity of a gentleman's narration" (Shapin 1994:83). The constraint on membership was designed to ensure that inquiry would be unbiased. Craftsmen, midwives, apothecaries, foreigners of all ranks, and women—even gentlewomen—were excluded because they were considered untrustworthy. Being dependent on others, they needed to curry favor. Thus, it was believed, they had incentives to lie. Members of the Society held themselves accountable to one another; their social status underwrote the expectation that they would conduct themselves and their research appropriately. At its inception, the Royal Society seems to have had been structured as a quirky gentlemen's club whose members shared an interest in investigating scientific questions.

Given the good of the practice—achieving an understanding of the natural world—the rationale for restricting membership was problematic for several reasons. First, the founders seem to have been rather unimaginative about motivations for lying, or at least about motivations for lying about one's scientific research. A scientist, even a gentleman scientist, might easily have any number of incentives to lie—for example, to feed his ego, best his rivals, enhance his reputation, alleviate his boredom, or promote a favored cause. Second, the emphasis on the danger of lying was unduly narrow. Lies are not the only way, and probably not the main way, that falsehoods get incorporated into science. At the cutting edge of inquiry, methods and techniques tend to be relatively crude, poorly aligned with their targets, and untested. Honest mistakes are common. This was something the Society failed to appreciate. Moreover, because a gentleman's word is his bond, members of the Royal Society were reluctant to challenge one another. Whatever a member reported about how he conducted his investigations and what he found was deemed to be acceptable. There was no incentive to replicate results or to critique methods. Indeed, attempting to replicate a colleague's finding was likely to seem a bit presumptuous, since it would implicate that he was not to be trusted. Nevertheless, automatically taking an investigator at his word is apt to entrench errors and impede scientific progress. Gentlemanly scruples thus interfered with devising appropriate norms of acceptance, methods of inquiry, and ways to identify and correct errors. Third, by excluding non-gentlemen, the Society deprived itself of the insights and abilities of instrument makers, navigators, apothecaries, and technicians who were apt to have considerable expertise in matters under investigation. This unduly limited their epistemic range. Initially, the goods in the practice were not well aligned with the good of the practice—understanding the natural world.

Over time the Royal Society changed its ways. It admitted commoners (initially under the aegis of a gentleman) and recognized that results had to be replicable if they were to be scientifically acceptable. It devised methods for identifying and rooting out received errors. It went on to formulate standards of acceptability more suited to promoting its epistemic end. The Society was, and still is, an epistemic community whose members shared a conception of what their end is, what means were appropriate, what obstacles might impede progress, and how to get around those obstacles. The initial arrangement fell short. So it introduced modifications, both in its membership criteria and in methodology, to bring its practices into better alignment with its goals.

Process reliabilists such as Goldberg (2010) might insist that the Royal Society was not an epistemic community until it developed and deployed reliable methods for generating results, where a reliable method is one that delivers truths but no falsehoods a high proportion of the time. I disagree. I think that from its inception the Royal Society was an epistemic community—a realm of epistemic ends. It began with a manifestly epistemic goal: advancement of the understanding of nature. It also had a conception of the impediments that stood in the way. The ill-advised exclusion of those who might be swayed by financial, social, or ideological pressures was fueled by the belief that those pressures were the forces that most needed to be blocked. It took an unduly dim view of women, foreigners, and the lower classes, and an unduly optimistic view of how easy it would be for unfettered inquiry to yield the sort of understanding it sought. With the growth of science came an increased appreciation of other factors that promote or inhibit the advancement of scientific understanding. The Society thus modified its practices and membership criteria accordingly. With advancement in the understanding of nature came advancement in the understanding of what it takes to understand nature. Standards for validating methods and verifying findings emerged.

My disagreement with process reliabilism might seem to be little more than a terminological dispute. Should we withhold the term "epistemic" until a process passes a certain threshold for truth conduciveness? Or should we count a practice and its products as epistemic when it (non-accidentally) advances understanding, even if the threshold is not reached?

In my view, a practice can non-accidentally advance understanding, even if its track record is far from stellar. Think, for example, of inquiries into dark matter. Because dark matter apparently does not absorb, emit, or reflect electromagnetic radiation, it is, as far as we can tell, unobservable. Astrophysicists infer that it exists because a variety of gravitational effects, such as the formation and evolution of galaxies, cannot be explained by current theories of gravity unless the universe contains vastly more matter than we can detect. But they do not know what dark matter is or even what it could be. They have eliminated the plausible and not-entirely-implausible candidates

Realms of Epistemic Ends 129

and have demonstrated that tweaking current theories of gravity is no help. They are far from having an acceptable account. Nevertheless, I maintain, in positing the existence of dark matter, they advanced understanding. Process reliabilists would presumably deny their inquiries epistemic status until they have an account that yields truths, but no falsehoods, in a high proportion of cases. This strikes me as unduly demanding. With each failed attempt, they make epistemic progress. They understand gravity a bit better as a result of their investigations. In any case, astrophysical models involve idealizations and simplifications that render them literally false. Some advance understanding by yielding false results that are not far from the truth; others, by yielding results that, although far from the truth, diverge from the truth in irrelevant respects (see Elgin 2017:23–32). The idea that we will ever have a theory of dark matter or even of ordinary matter that does not involve idealizations and simplifications is unrealistic.

The suggestion that the dispute concerns thresholds is off the mark. An epistemic community devises practices to promote its ends—perhaps understanding, prediction, and control of nature or of galaxies or of traffic patterns. Conceptions of evidence, measurement, validation, and confirmation are defined within a practice—by standards a community sets for itself because it believes that by satisfying those standards, it will promote the goods of the practice. Only to the extent that the community realizes the goods in the practice has it any reasonable hope of achieving via the practice the good of the practice.

Epistemic Communities and Their Ends

What qualifies a community as "epistemic"? It is not enough that a community purports to pursue truth and devises practices with that goal in mind. Communities of astrologists, psychics, and conspiracy theorists do that. But if I deny that truth conduciveness or an analogous success metric is required for epistemic standing, how can I exclude them?

An epistemic community is constituted by joint commitments about the ends and means, standards, principles, and priorities of its epistemic enterprise. This makes it a systematic union of different rational beings through common laws. Very roughly, members are unified in agreeing what they want to understand, how to go about finding it out or figuring it out, and why their methods are appropriate, given their epistemic ends. There are a variety of epistemic communities whose interests, priorities, standards, methods, and goals diverge. A community interested in the life cycle of the gypsy moth has different specific goals, different priorities, and different methods from one interested in the development of the sonata. The methods suited to one would be ill-equipped to foster the ends of the other. Even communities that share a common domain—for example, molecular biologists and ecologists—may differ in their priorities, methods, and specific goals. This way of putting it may make things look overly intellectual—perhaps more suited to high-level disciplinary understanding than to everyday epistemic activities. I disagree. Serious football fans constitute an epistemic community. They analyze players and plays, strategies and schedules, coaching and officiating—even the weather—all with the goal of understanding what accounts for last week's debacle. They dismiss with contempt the simple (albeit correct and uncontroversial) explanation that their team lost because, at the end of the game, the opponent had a higher score. They know, and know that one another know that. But that's not the sort of understanding they are looking for.

Despite significant divergences in ends and means, epistemic communities share generic requirements and face common constraints. Among the universal requirements are logical consistency, internal coherence, and responsiveness to reasons. One general methodological requirement on epistemic practices is that like cases be treated alike. Practices diverge over what makes cases relevantly alike and what qualifies as treating them alike. However these issues are settled, the demand for sameness of treatment immediately follows. If a and b are alike in every relevant respect, then to treat them differently would be arbitrary.

Arbitrariness is not randomness. Very roughly, a random process is one where there is no telling in advance what, among specified alternatives, the outcome will be. Often there are good reasons to accommodate and exploit randomness. Some random processes, such as radioactive decay, are instances of indeterminacy. There is no fact of the matter as to exactly what will happen until the process occurs. Other processes have foreordained outcomes given the laws of nature and the initial conditions, but the situation is so complicated that figuring the result out in advance is hopeless. Coin flips are random for this reason. Because we are incapable of computing in real time all of the forces acting on the coin, we are ignorant of the outcome until the coin lands. In yet other cases, the role of randomness is to reflect not ignorance but indifference. When a sleight-of-hand artist invites you to pick a card, any

Realms of Epistemic Ends 131

card, he does so because the trick will work regardless of which card is chosen. Similarly, some mathematical proofs show that something holds generally because it holds for a randomly chosen case. The proof that the square of any odd number is odd proceeds by choosing an odd number at random and showing that its square is odd. For the purposes at hand, it simply does not matter which alternative is chosen. Because accommodating indeterminacy or ignorance may be unavoidable, and acknowledging and exploiting indifference can be epistemically fruitful, there is no bar in principle to incorporating randomness into epistemic practices.

Arbitrariness, as I use the term, is a matter of favoritism or caprice. Rather than basing a choice on the merits of the alternatives or resorting to a random process if the alternatives are equally choice worthy, choosing arbitrarily imports irrelevant factors to tip the scales. If a and b are alike in all relevant respects and one must be chosen, it would be arbitrary to favor a over b on the irrelevant grounds that vowels are more likable than consonants. If b is preferable to a on the basis of relevant considerations, it would be arbitrary to choose a over b by favoring irrelevant ones. An epistemic practice is designed to achieve an epistemic goal. To incorporate grounds for favoritism that distort the evidence or the weight of evidence is to subvert the practice, undermining its prospects of achieving the goal.

A standard objection to conspiracy theories is that they use different metrics to assess friends and foes, even though the conduct under assessment is the same. Such assessments are epistemically untenable because they are arbitrary. They import and rely on irrelevant factors. Granted, it is often difficult to determine which cases are alike. Criteria of relevance are crucial. But the standards must be uniform and non-question begging. They cannot permissibly be gerrymandered to yield the results one wants.

Practices operate not in a vacuum but in natural, social, and normative environments that affect what they can achieve and how they can achieve it. To be effective, their design must be sensitive to constraining conditions. A practice that calls for a perpetual motion machine is doomed from the start. Laws of nature are inviolable. Local conditions that are not prohibitive nonetheless set constraints. In northern New England, the Connecticut River is apt to be frozen until the end of March. The Dartmouth rowing team's spring training regimen must, in one way or another, accommodate itself to that fact. The expected thaw date is as much a fact for the college choir as it is for the rowing team. But it is a fact that has, and should have, no effect on the ways the choir pursues its ends. For the oarsmen, however, the expected thaw date is not just a brute fact. It is something that constrains their pursuit of their ends. They need to frame their training around it. Sensitivity to relevant environmental conditions is mandatory.

Our epistemic practices are bounded by two endemic human limitations: (1) human beings are fallible—despite our best efforts, we make mistakes; and (2) human beings are not omniscient—despite our best efforts, there remain facts we are ignorant of. These are practically platitudes, but they are not idle. Our practices need to accommodate these limitations, not just acknowledge them. We require norms of accommodation that incorporate an appreciation of our inevitable limitations.[2]

It follows from our fallibility that regardless of the weight of evidence that supports a contention, it still might be wrong. When a community purports to pursue understanding but exempts certain contentions from scrutiny, it treats those contentions as incorrigible and reliance on them as infallible. If such a contention is in error, the community's practices will, by design, be powerless to discover it. To treat as irrevocable a contention that might be wrong is to invite the entrenchment of error. Epistemic practices should treat acceptance as provisional and revocable.

The phrase "treat as" is important here. It connects a contention with the epistemic agents who use it and constrains the uses they can permissibly make of it. To treat a contention's acceptance as provisional is to incorporate it into one's network of commitments in such a way that it can be rescinded should considerations emerge that discredit it. A potentially erroneous contention should not be accepted dogmatically. If it might be called into question, the evidence that supports it should not be thrown out. If it is challenged, the practice should have resources to reevaluate it. Without the evidence, it would be ill-equipped to do so.

For a contention to be treated as revocable, there should be mechanisms for revocation—ways to excise it from the epistemic network it belongs to or to alter its contribution to that network. Newspapers issue corrections for mistaken statements of fact; scientific journals retract repudiated or irreplicable findings. Explicit retractions are not always possible, nor are they always effective. The more intricately a commitment is entwined in the fabric of understanding, the harder it may be to untangle. Nevertheless, it must be possible to rescind or revise our commitment to it. Although, for example, Newtonian

Realms of Epistemic Ends

assumptions are deeply interwoven into everyday practices, scrupulous scientists and science teachers take pains to frame them as approximations or idealizations rather than as the unvarnished truth. Rather than extracting a commitment, a practice may reconstrue it, indicating that it stands in a different relation to the facts than we might once have thought and might still naïvely be tempted to think. It is, for example, a model, an idealization, or an approximation rather than, as one might have thought, an exact literal truth. This is a familiar and valuable epistemic strategy. It enables us to retain viable aspects of a commitment while sifting out the dross. To use this strategy responsibly requires recognizing that the commitment as currently construed does not function as a literal truth.

Astrology and parapsychology privilege certain commitments, protecting them from refutation. Evidence cannot discredit them, nor can clashes with the findings of other disciplines. Even logical inconsistencies are brushed aside. Their fundamental claims are taken to be inviolable. That being so, their practices do not qualify as epistemic.

Accommodation to fallibility should do more than concede that a contention is incorrect when evidence of error is overwhelming. Epistemic practices should be sensitive and responsive to intimations of error. An untoward observation, a putatively plausible analogy that seems strained, or a cluster of data points which although strictly consonant with an accepted contention seems surprising in light of it—these are the sorts of things that might and sometimes should give us pause, triggering a reconsideration of an accepted claim, method, or standard.

An epistemic practice should provide forums for raising challenges and safeguards that ensure that such challenges are given their due. Hospital morbidity and mortality conferences analyze adverse patient outcomes in an attempt to discover what, if anything, went wrong. By rigorously examining the sequence of events that led to the regrettable outcome, as well as the procedures, assumptions, and policies that were in effect, physicians often identify medical errors and devise modifications to prevent their recurrence. Although such conferences are useful for identifying the sources of one-off errors, they are particularly valuable because they often uncover systematic oversights—aspects of standard practice that leave the door open to further unfortunate outcomes. Not every epistemic practice has the sort of formal, systematic review that is characteristic of morbidity and mortality conferences. Many of the mechanisms for rooting out received errors are informal and opportunistic. But for a practice to accommodate human fallibility, it should be sensitive and responsive to the possibility of error.

Because we are not omniscient, epistemic practices ought not be insular. An epistemic community should be prepared to entertain insights drawn from the outside, capable of recognizing their relevance, and flexible enough to incorporate them into its practice if they are sound and look to be fruitful. A practice's standards of evidence and relevance must therefore be public, open to criticism, and susceptible to revision. There should be mechanisms for determining when revision is called for. Moreover, to accommodate our lack of omniscience, the community that constitutes a practice should be prepared to revise its membership. It should be open to recognizing that epistemic agents who are currently considered ineligible for membership may have something of value to offer. The evolution of the Royal Society provides a case in point. Early on, it became evident that relevant expertise was to be found among women, foreigners, and commoners, and that incentives to lie about scientific matters were independent of pedigree. As a result, the Society gradually welcomed a wider range of investigators and introduced checks on findings, recognizing that doing so expanded its epistemic range. It remains an elite society; its intellectual standards remain high. But (apart from the automatic inclusion of members of the British royal family), social considerations are irrelevant.

An epistemic community has an incentive to avail itself of ideas, methods, and techniques that relevantly expert outsiders possess. To exclude such outsiders would be epistemically unjust, effectively silencing them by prohibiting them a hearing (see Fricker 2009). It would also epistemically impoverish the community that excludes them, blocking access to valuable perspectives and insights. When doctors ignore nurses, professors ignore teaching fellows, or craftsmen ignore apprentices, valuable information is apt to be lost.

Overlooking or underrating the expertise of those considered unqualified can be costly, as this tragic episode shows. In 1959, the Società Adriatica di Elettricità completed construction of a dam in the Vajont Valley in Northern Italy.[3] Before building the dam, geologists did a site study that relied heavily on general knowledge about the sorts of rocks—mainly limestone—that comprised the terrestrial crust in the region. According to the head geologist, "Overall, limestone is honest, because it reveals its flaws on its surface" (Barrotta and Montuschi 2018:21). Since no serious surface flaws were found,

Realms of Epistemic Ends

they inferred that the valley was geologically stable. Having deployed both theoretical and material models to test their assumptions, they concluded that the project posed no threat. Nevertheless, there was considerable local resistance to the project. Inhabitants of the region vehemently disagreed with the scientific findings. Although the scientists, engineers, and policymakers made some efforts to placate the resistors, they failed to give the locals' worries their due. On October 9, 1963, a landslide plunged 250 million cubic meters of debris into the reservoir behind the dam, giving rise to a massive wave that flooded the valley, killing more than two thousand people.

Although the geologists had considerable general understanding of the behavior of limestone, they lacked specific knowledge of the limestone in the Vajont Valley. Nor evidently, did they think they needed it. The denizens of the valley, for their part, probably lacked general geological understanding, but as their families had lived in the region for generations, they had plenty of specific local knowledge. They were well aware that the area was susceptible to landslides. Whatever may be true of limestone in general, inhabitants knew that the rock crust in their region was unstable. Although they tried to make their case, their worries were sidelined.[4]

We can see why. The locals were not scientifically sophisticated. They did not speak the language of the academy, nor could they support their claims by statistical analyses or scientific models. Their evidence was anecdotal; it seemingly did not generalize. Their objections did not, that is, satisfy the standards of acceptability of the scientific community. As a result, they were victims of epistemic injustice (see Fricker 2009). They could not get uptake for their claims. They ought to have been heard and taken seriously. They had relevant, expert knowledge about their particular corner of the world. Minimally, the inhabitants' claims should have prompted further investigation into the geology of the region. It should have served as a reminder that what is generally the case is not always the case and that, since models make contrary to fact assumptions, results ought not automatically be read back onto the phenomena they bear on. I will discuss the issue in greater detail in chapter 11. Moreover, the clash between the well-established general claims about the honesty of limestone and the residents' insistence that local limestone had a dodgier profile should have prompted greater in situ investigation. These are considerations that the geologists, by their own standards, ought to have recognized. The difficulty is that the objections that should have moved them did not satisfy scientific standards. To take the residents' input seriously, the geologists would have had to relax their commitments to rigor, to sources of evidence, and to generalizability in order to entertain the possibility that the local critics had important information that their commitments blinded them to.

Several important epistemological lessons emerge. As we saw with the evolution of the Royal Society, expertise is widely distributed; it is not a matter of pedigree. Moreover, there may be valuable local expertise that does not generalize. The commitments that structure an epistemic community at any given time both enhance and limit understanding. At the frontier—across disciplines or between disciplinary and less regimented understanding—standards need to be flexible so that parties can take advantage of their interlocutors' expertise (see Nersessian 2022). Principles of charity may need to be invoked to figure out, across conceptual, terminological, and methodological divides, what the others are thinking and why, or if, it should be accommodated.

Even so, a community ought not open its doors to crackpots. To do so would swamp it with useless and baseless information. It does no good and potentially much harm to devote resources to entertaining and refuting the ideas of those who think that the stock market is controlled by space aliens or that vaccines implant microchips.

Whether and how to expand an epistemic community depends on a variety of factors, including an assessment of who, beyond the current members, might have useful insights to offer. An epistemic community does itself a disservice if it is unduly restrictive or skewed in its membership requirements. Its membership criteria must promote its epistemic ends. It is reasonable to exclude those who lack the requisite expertise. But the excluded must actually lack that expertise. It is not enough that they lack its social markers—that candidates for membership are not gentlemen or did not attend the right schools or lack the appropriate degrees or fail to look the part or speak the language. We have to use proxies. We cannot peer into a person's head and see whether she knows quantum electrodynamics or auto mechanics or Aristotelian metaphysics or limestone geology. To best serve their epistemic ends, communities of inquiry should continually vet and update their proxies and their grounds for trusting them. They should regularly rethink what kind of expertise they need and where it is to be found. As their understanding of a topic grows, so should their understanding of whose expertise might be valuable. And as that grows, the membership criteria should evolve accordingly.

To accommodate our limitations, epistemic practices need first to recognize them—to refrain from overgeneralizing or unduly restricting, to recognize

Realms of Epistemic Ends

margins of error, to discern blind spots. They also do well to devise methods and mechanisms for expanding our epistemic range—material devices like lenses, scanners, and probes; conceptual devices like mathematical techniques; social arrangements like interdisciplinary collaboration.

Political Relations

I said that epistemic communities justifiably restrict their membership to those who have the requisite expertise, and I argued that they ought to be open to finding that expertise in unexpected places. But I have not yet discussed the internal structure of epistemic communities—the relations in which members should stand to each other. To best promote its epistemic ends, a community should be so structured that in their joint epistemic activities its members are politically free and equal.[5] They should be free to venture any hypothesis or recommend any method they consider potentially worthy of community endorsement. They should be equal in their right to have their hypothesis or method seriously entertained. It is not enough that they get to put their proposal on the table. The community owes it to them and to itself to give the proposal a fair and open-minded hearing. That is, there must be uptake (see Longino 1990:66–81). There is, of course, no assumption that every proposal is equally worthy of endorsement. To give a dumb idea its due is to seriously entertain it and soundly reject it.

Although the requirement is phrased as a *right* to equal treatment, it is a right that imposes a reciprocal duty. Members of the community owe it to one another to venture only hypotheses that there is reason to think merit the community's consideration. The community would be impeded in its pursuit of its epistemic ends if the members did not take this responsibility seriously. Thus, one obligation incumbent on autonomous epistemic agents is to create and sustain a community where the pursuit of understanding can thrive.

The social structure of an epistemic community is more intricate than the political requirement suggests. Members of a community differ in levels of expertise. Some are more skilled or more knowledgeable than others. Although all have an equal right to venture hypotheses or methods and have them taken seriously, it is no surprise that some do better than others in having their proposals accepted. This is as it should be, so long and only so long as the proposals are assessed on their merits rather than on their pedigree.

A complicating factor is that members of the same community often differ in areas of expertise as well. My earlier discussion might have suggested that the support an individual epistemic agent gets from his compatriots just strengthens the justification he generates on his own. He took five readings; they added another ten. Often, however, an epistemic agent relies on his compatriots for skills he does not have or information he cannot personally access. He never took statistics or can't read Greek or doesn't know how to interpret a Feynman diagram. So he turns to others. The division of epistemic labor enables him to augment his epistemic reach by drawing on their expertise. This is an important dimension in the community's being a *systematic* union of rational agents. There is epistemic value in the fact that the members of the union are not all alike.

Although the division of epistemic labor is valuable, it makes those who rely on it vulnerable. Where they differ in expertise, agents cannot directly assess one another's reasons or check their findings. The bench scientist may have no clue why the statistician ran the regression she did; the historian, no sense of whether the classicist's translation is accurate; the chemist, no idea whether the physicist's reading of the Feynman diagram is correct. To some extent, individual epistemic agents can finesse this worry by appealing to multiple experts, expecting that if the report is tenable, consensus will emerge. But ultimately, it is a question of trust. Members of an epistemic community need to trust those whose expertise they draw on. Hence the experts need to be trustworthy. And given the free and equal requirement, any member whose proposal is reflectively endorsed becomes, at least pro tem, an expert vis-à-vis the content of that proposal.

Trustworthiness is required of every member of the community. Some epistemic communities are large and dispersed across the globe. Members are apt to be strangers to one another. Thus their trust in each other must be impersonal. The community needs to forge institutional bonds of trust that make it reasonable for its members to rely on one another, even when they are strangers to one another. There are then moral threads in the tapestry of epistemic commitments that binds a community of inquirers.

I have argued that epistemic autonomy and epistemic interdependence are mutually supportive. Autonomous agents join together to create communities within which they can pursue their inquiries as legislating members of realms of epistemic ends. Earlier I focused on the individual agent, arguing that, given our human limitations, we must rely on one another. We do so in ways that do not compromise but rather leverage our autonomy. As legislating members of realms of epistemic ends, we make the rules that bind us, and we reflectively endorse them and the practices they constitute. In this chapter, I focused more on the structure of an epistemic community, the limitations it must accommodate, and the goods that it generates. Working together, autonomous agents establish epistemic goals, and devise and regularly revise practices designed to realize those goals. As understanding evolves, they revise the goals as well.

This, no doubt, is an admirable human achievement. But it leaves an important question unaddressed: Why should we think this will work? We try very hard to design practices that will promote our epistemic ends. We take one another's views seriously and come up with results we are willing to stand behind because they satisfy the standards we collectively set. We seek to understand the way the world is, which is largely independent of what we think and do. What guarantee do we have that by satisfying our own standards, we achieve our epistemic objectives? Answer: there is no guarantee. We can continually improve our take on things by extending and deepening our understanding, thereby uncovering errors, gulfs, and blind spots. But we cannot perfect it. If we have woven a dense tapestry of commitments in reflective equilibrium, we are in a position to say that if we are wrong about this, we're wrong about a lot. But, of course, we might be wrong about a lot. Maybe we will wake up on Judgment Day and discover that we were brains in a vat after all.

We design our epistemic practices to be self-monitoring and self-correcting. We open our learned societies to agents whose expertise we now recognize but originally doubted. We revise our experimental practices to introduce suitable controls. We fine-tune our logics to discredit inferences we deem invalid. We submit findings to peer review. We have plenty of evidence that our understanding of the world is better than that of our forebears. But whether we will ever arrive at a complete and accurate understanding of the phenomena is unknown.

I rather hope we will not. Peirce (1986), Sellars (1968), and Putnam (1978) maintain that inquiry is converging on the truth. The end of inquiry, they maintain, is the finish line that inquirers are racing toward. Dewey (1916) has a different view. Rather than converging, he believes, effective inquiry branches out. In discovering the structure of DNA, Watson and Crick not only solved an outstanding problem they also set the stage for a new science: molecular genetics. New questions could be framed, new techniques devised, new understandings achieved. There is no end in sight.

6 Word Giving

Testimony

Testimony is a vehicle for efficiently and systematically acquiring information that we cannot easily obtain firsthand. In the prototypical case, an informant imparts information to an audience with the aim of having it accepted. An act of testifying is successful only if the audience accepts the information as a result of the informant's intentionally imparting it. So far, this concerns only the cause of coming to accept. That is epistemically inert. But testimony is epistemically significant because the testifier not only imparts information she also makes it reasonable for the audience to credit that information. Testimony conveys warrant. It is a source of epistemic entitlement, because the testifier vouches for the information she imparts. If she is to be trusted, then unless they have defeaters—that is, unless they accept considerations that discredit the information or the warrant for it—the audience can come to know (or at least be justified in accepting) the information thus imparted.

The structure of the testimonial exchange should be spelled out more fully. Let S be the testifier, H be the recipient, and p be an informative statement. In an epistemically felicitous case of testimony:

(1) S knows that p;

(2) S presents herself as someone who knows whether p;

(3) S asserts that p;

(4) S intends that H accept p in consequence of her asserting that p;

(5) H hears or reads S's assertion;

(6) H correctly grasps the content and force of the assertion (That is, he both understands the content and understands that the utterance or inscription is an assertion rather than some other speech act);

142   Chapter 6

(7) *H* recognizes that *S* is presenting herself as someone who knows whether *p*;

(8) *H* takes *S*'s assertion as sufficient reason to accept *p*.

Then, since *p* is true (which follows from the fact that *S* knows that *p*):

if *H* has no defeaters and comes to accept *p* on the basis of *S*'s say-so, *H* knows that *p*.

In that case, *H*'s knowledge that *p* is grounded in the fact that he properly uptakes the information that *p* from a knower.

Giving directions is often presented as a paradigmatic case of conveying knowledge via testimony. Suppose Sally tells Harry that Alewife Station is on the Red Line. She is, and presents herself as, an epistemically competent informant. She supplies the information in a way that makes it plain that she takes herself to know what she is talking about. Grasping the content and force of her utterance—that is, appreciating that her utterance is an assertion to the effect that Alewife is on the Red Line—Harry takes her at her word. Since there is nothing sinister about the encounter and nothing tentative about Sally's utterance—nothing, that is, that would or should lead Harry to fear being misled—her assertion gives him good reason to believe that Alewife is on the Red Line. Her assertion is true, and he has no defeaters. So he knows that Alewife is on the Red Line. Testimony like this is an everyday means of coming to know.

Several aspects of this exchange are worth noting. First, Harry's acquisition of knowledge here is not cognitively costly. He did not have to do much to gain epistemic entitlement to the information Sally conveyed. Nor did Sally have to do much to convey the information and the epistemic entitlement to it.[1] Perhaps significant effort was required for the original source of the information to come by it in the first place. But the transfer from Sally to Harry was easy. Second, the epistemic affordances Harry gains are fairly narrowly circumscribed. He can use the very bit of information he gleaned with confidence. If he wants to go to Alewife, he can safely hop on the Red Line. He can responsibly transmit the information to another inquirer. He can draw trivial inferences from it. His knowledge that Alewife is on the Red Line is secure, even if he knows virtually nothing else about local geography or public transportation. To be sure, he might know a lot more. And the new information may be valuable to him precisely because it fills an irksome gap in his overall understanding of the Boston subway system. But he knows that Alewife is on the Red Line even if that knowledge makes a minimal contribution to his cognitive economy. It is fully in order even if it is just an isolated bit of knowledge. Third, when asked "How do you know?" he can and should answer, "Sally told me." Apart perhaps from his lack of defeaters, nothing more is required for this answer to be epistemically adequate.

We might modify the schema slightly. Rather than insisting that the informant assert, it is enough if she professes. When Sally says that the trip to Alewife will take twenty minutes, Harry recognizes that that is a ballpark figure. It is true enough but probably not exactly true. When he boards the train, he accepts what she says. It is a good enough basis for inferences and actions relevant to his getting to his destination on time. It makes no difference that testimony involves speaking loosely, so long as all parties are attuned to how loose it is.

To say that the affordances of testimonial knowledge are narrowly circumscribed is not to suggest that the content conveyed by testimony need be sparse or simple. Testimony is not restricted to conveying isolated factoids. The content of a testimonial utterance or inscription may be extended and complicated. Tattletale Timmy testifies that Maya did not eat her spinach and Billy did not drink his milk and José did not shine his shoes and. . . . The content of Timmy's testimony is logically complex, having the form of a conjunction: $a \mathrel{\&} b \mathrel{\&} c \mathrel{\&} \ldots$. Typically, in giving directions, spatial and perhaps temporal order is significant as well. Sue tells Marty that to get to the store, he should go straight for four blocks, then go left for two. The order matters. The recipient of her testimony should go straight prior to turning left. Otherwise, he will end up in a mystifying maze of alleys. Here the form is: *a before b*. In yet other cases, testimony conveys dependence relations (see Greco 2020). The fire marshal testifies that the fire was caused by faulty wiring (Pritchard 2010:81). Here the form of the testimonial content is: *a because b*.

Nor, of course, is there any limit on the length of testimony. Via testimony, an engineer could, for example, describe exactly how a complicated machine works. The testimony, perhaps including gestures and diagrams, is far from straightforward. Nevertheless, the testimony is circumscribed in that the extent to which it would equip and entitle the hearer to go beyond the information it contains may be severely limited. His testimony-based knowledge would be fully in order if, because he had been told, he knew that the value displayed on a certain monitor should stay between 630 and 760. But the testimony need not give him any idea what would happen if the value exceeded 760 or why its staying within the prescribed limits matters. Nor need it provide knowledge of what magnitudes the various dials measured, why their measurements were important, or how they or the mechanisms they were connected to worked.

Inasmuch as testimony may be long and complicated, the claim that it is narrowly circumscribed may seem surprising. But the limit in question is not a matter of constraining the breadth or complexity of the information conveyed. It is a matter of restricting the audience's epistemic resources for going beyond that information.

Such circumscribed knowledge is epistemically valuable. We may not want or need or be able to find something out for ourselves or be in a position to personally supply substantive justification for facts we seek to know. Often we have no incentive to go beyond the information given. Then our cognitive and practical goals are met by an isolated bit of reliable information. Testimonial knowledge plays a vital role in the division of cognitive labor.

Acquiring a Right

Testimony is a mode of word giving. People continually impart information to one another. "Take my word for it," they urge. "You can count on me." Nevertheless, it is not altogether clear what it is to take someone's word or when it is reasonable to do so. Judith Jarvis Thomson advocates accepting:

> *The Assertion Thesis*: Y gives X his or her word that a proposition is true if and only if Y asserts that proposition to X, and
>
> (i) in so doing Y is inviting X to rely on its truth, and
>
> (ii) X receives and accepts the invitation (there is uptake). (1990:298)

If the Assertion Thesis is correct, the word giver issues an invitation; the word taker accepts it, thereby acquiring a right. In particular, she acquires a claim against the word giver—a claim that is infringed if the proposition in question is not true. I'd modify this to say that the claim is infringed if the proposition is not true enough. A little roughness around the edges typically does not amount to an infringement. Thomson focuses on promising, where the moral dimension of word giving is particularly salient. But she recognizes that there are other modes of word giving as well. I use her account as a springboard to shed light on the epistemology of testimony.

Word Giving 145

Testimony is a mechanism for information transfer. The guide says, "The cave paintings at Les Eyzies are seventeen thousand years old." The reporter announces, "The Dow lost 180 points today on heavy trading." The physician warns, "Hypertension increases the risk of heart disease." The passerby obliges with directions, "The museum is two blocks down, on the left." In each case, the speaker represents herself as qualified to speak knowledgeably. Although she intimates that her assertion is backed by epistemically adequate reasons, she does not supply them. Testimony then conveys information without supplying arguments or evidence to back it up. To be sure, an idle assertive aside could do that. But because testimony is a mode of word giving, it does more. The testifier invites her word receiver to accept what she says on the basis of her say-so. She warrants that her testimony is acceptable. She implicates that it is backed by what he would consider good reasons—that it satisfies Grice's evidence maxim and would be endorsed by members of his relevant community of epistemic ends.

This raises a question: Is testimony any different from mere assertion or profession? In asserting or professing that p, the speaker implicates that she has satisfied the evidence maxim. That is, she implicates that her claim is backed by adequate reasons. Her utterance, it seems, already comes with a guarantee. The difference cannot consist in the fact that testimony conveys warrant. Rather, the difference lies in the fact that testimony is directed. It imparts information *to* someone or some group. It invites those to whom it is addressed to accept the information, and it gives them a claim against the testifier, should the testimony turn out to be unacceptable. An eavesdropper, who acquires the same information and recognizes that its utterance implicates warrant, has no such claim. He is on his own. Should a speaker's testimony turn out to be unacceptable, she will have done her audience but not her eavesdropper a wrong.

If we understand the nature of that wrong, we get a handle on what the good of testimony is, what benefits it provides. Here the contrast with promising is helpful. Promising provides a framework for voluntarily restricting one's freedom. It facilitates planning and fosters cooperation. Thomson identifies several characteristics of the word giving that constitutes promising. (1) Promising is future directed. The propositions whose acceptability a promisor commits herself to are in the future tense. I can promise that I will eat my spinach. But if I assert that I am now eating my spinach or that I ate my

146   Chapter 6

spinach yesterday, my assertion is not a case of promising. (2) Promising
has the promisor as its subject. I can promise that I will eat my spinach. I
can promise that I will do my best to get Sam to eat his spinach. But I cannot
promise that he will eat his spinach. The reason, evidently, is that no act or
omission of mine can ensure his compliance. Promising then is essentially
first-personal.[2] (3) Only a limited range of actions, refrainings, or states of
affairs fall within the scope of promising. I cannot promise that I will live
to be 150, for I lack, and am well aware that I lack, the capacity to bring
that about (Thomson 1990:299–300). Taken together, these features show
that promising is restricted to future contingents that are within the actor's
power. To the extent that it is indeterminate which states of affairs are contingent in the relevant sense and which of those are within an actor's power,
the scope of promising is indeterminate as well.

Testimony consists of statements of purportedly established fact. It has
no restrictions as to tense or person. I can testify that Woodrow Wilson was
president of Princeton University, that $E=mc^2$, that I am a resident of Massachusetts. I can't testify that I will eat my spinach though, for despite my
best intentions, I might not. Future contingents then lie outside the scope of
testimony. But not all statements about the future are excluded. If a prediction is so firmly grounded in established facts and laws that its truth is not
up for grabs, it can be the content of testimony. A scientist can testify that a
sample of plutonium will continue to emit radiation for hundreds of thou-
sands of years, since well-established physical facts and laws underwrite that
prediction. There may, of course, be some question as to what facts and laws
are capable of underwriting testimony about the future. So whether a particular prediction qualifies as testimony may be controversial. But a statement's
being in the future tense does not automatically rule it out.

Talk of future contingents and freedom to act is apt to induce flutters of
metaphysical anxiety. Is the future genuinely open? Is it open in the ways
that we think it is? Do we even have a clear conception of what it means to
say that it is? Are human beings genuinely free to choose and able to act as
they choose? Are we free and able in the ways that we think we are? These
are legitimate questions whose answers are by no means obvious. If we have
to answer them correctly in order to explicate word giving, our prospects
are bleak. Luckily, we need not do anything so ambitious. Promising, testimony, and other modes of word giving are human practices. They depend
for their utility not on what is *really* the case with regard to contingency or

Word Giving

human freedom in some metaphysically robust sense of *really* but on shared assumptions about these matters. Even if human beings can, through a sheer act of will, live to be 150 years old, no one thinks that we can do this. So we are unwilling to make or to accept a promise to live that long. Even if a neonatologist's predictions about infants' eventual career choices have as high an objective probability as physicists' predictions about radioactive decay, we do not consider psychological predictions anywhere near that good. So a responsible neonatologist would not proffer, nor would we accept, such a prediction as testimony. Word giving is circumscribed by shared, commonsensical assumptions about metaphysical matters. Many of these assumptions are vague and inarticulate. Some, no doubt, are false. But because they are shared, they supply the mutual understanding that we need for the issuing and accepting of invitations to take someone's word.

To explicate word giving, we need to recognize the shared assumptions that underwrite them. We understand a good deal about practices when we see how those assumptions function. If everyone agrees that people in general have the ability to return books that they borrow, we permit one another to promise to return books and hold them responsible for their failures to do so. If everyone agrees that some people are cognitively competent to calculate rates of radioactive decay and to report the results of their calculations accurately, we count expert assertions about such matters as testimony and consider testifiers blameworthy if their reports are inaccurate. By reference to the presuppositions in effect then, we can make sense of the actions, motivations, and assessments they give rise to.

"Ought" implies "can." If a person cannot do p, he is under no obligation to do p and cannot rightly be faulted for failing to do p. Appeal to shared presuppositions explains why we hold people responsible when we do. But if the presuppositions are wildly off the mark, we may be holding people responsible when in fact they are not. Doubtless, we sometimes hold people responsible for things they could not avoid. Probably some of our mistakes are due to our faulty views about matters like freedom, agency, and contingency. Still, our word-giving practices are remarkably successful. People frequently behave in the ways they promised they would and acknowledge their wrongdoing when they fail to do what they promised. Knowledgeable informants often convey information that other evidence bears out and admit to their mistakes when the evidence shows them to be wrong. This indicates that however inaccurate the underlying assumptions are, they are not so wide

148   Chapter 6

of the mark that they discredit our word-giving practices entirely. I suggest then that we bracket concerns about metaphysical underpinnings and proceed on the assumption that our word-giving practices are reasonably well founded and do pretty much what we take them to do.

If I promise you that I will eat my spinach, I give you my word that "I will eat my spinach" is true. I give you a right to expect that I will eat my spinach. Of course, you already had *a* right to expect that. Freedom of thought ensures that you have a right to expect anything you like. You want to expect that I will eat my spinach? Who's going to stop you? But if an expectation grounded in nothing but freedom of thought is unfulfilled, no one is to blame. When I make a promise, the situation is different. I give you a claim against me. Ceteris paribus, if despite my promise, your expectation is unfulfilled, I am at fault. I gave you a reason to expect that I would eat my spinach, a reason that otherwise you would not have had. In giving you that reason, I warranted the expectation that I will eat my spinach. You have a claim against me then because I altered your epistemic circumstances. This epistemic element to promising is, I suggest, what converts the bare right into a claim.

A claim, Thomson argues, is a behavioral constraint. In giving you a claim against me, I agree to keep my behavior within particular bounds. In promising to eat my spinach, I agree to constrain my future behavior so as to include spinach consumption in it. The burden I shoulder is to make "I will eat my spinach" true. Plainly, I am up to the task. But I can testify to all sorts of things that I am utterly powerless to affect. I might, for example, testify that the cave paintings in Les Eyzies are seventeen thousand years old. Clearly, there is no way that I can make that statement true. The question arises: In so testifying, what claim do I give? How is my behavior constrained? If I don't eat my spinach when I promised that I would, I am subject to censure for failure to eat my spinach. But if the cave paintings are not seventeen thousand years old, it's hardly my fault. There is no way I can bring it about that the paintings are as old as I say they are. "Ought" implies "can." If I cannot make it the case that the paintings are seventeen thousand years old, I am under no obligation to do so and should not be faulted for failing to do so.

Nonetheless, I can be faulted. Why? Perhaps the most obvious answer is causal. My testimony that p caused you to accept p. So it might seem that I am to blame for your acceptance. But I can cause you to accept p in any number of innocent ways. You might, for example, overhear me rehearsing my lines for a play, mistake my utterance for an assertion, and so come to

Word Giving   149

accept what you take me to assert. Although my utterance of p caused your acceptance, the mistake is surely yours. I am not responsible for your misconstruing my speech act and acquiring a misconception as a result. Maybe a more complicated causal story is needed. Perhaps I am to blame for your wrongly accepting p if you come to accept p because you rightly think that I accept p. This is more plausible, but it still won't do. Suppose you overhear me sincerely saying that p and rightly conclude that I accept p. You therefore accept p on the basis of my assertion. What you don't realize, though, is that I am speaking to my therapist and that my assertion is (and, indeed, I recognize that it is) one of the baseless commitments that I am in therapy to overcome. Although I have plenty of evidence that $\sim p$, I cannot disabuse myself of my epistemic commitment to p, having been taught that p at a particularly impressionable age. Again, it seems that I am not at fault for your mistake. It is not enough that I cause you to accept p or even that I cause you to accept p by causing you to recognize that I accept p. I am responsible for your error, not when I cause you to accept p but when I entitle you to accept p. I convey to you not just a first-order epistemic commitment but also a right to that commitment. As in promising, you already have *a* right—a moral right to accept whatever you like. But neither that right alone nor that right in conjunction with a causal story of how you came by it gives you a claim against me. You have a claim against me because I invite you to take my word. I volunteer to shoulder the epistemic burden. Testimony, like promising, is a liability-shouldering device (Thomson 1990:94–95).

In testifying that p, I implicate that you can rely on me. For what? Let's look again at promising. When I break my promise, it is not because I failed to eat my spinach *simpliciter* that I am to blame. People are, in general, under no obligation to eat their spinach. I am to blame because I failed to eat my spinach *having given my word that I would*. Similarly, I am not to blame for the fact that the cave paintings are not seventeen thousand years old, but I am to blame for unjustifiably warranting your acceptance *that they are*. In both cases, it seems, what is at issue is a conjunction of the form:

p & Y gives her word to X that p.

The promisor can affect the acceptability of each conjunct. She can either bring it about that p or she can refrain from giving her word that p. The testifier can affect only the second. So the locus of blame may be different. The promise breaker is subject to reproach for failing to keep her word. In

150   Chapter 6

the case of testimony, there is no question of keeping one's word. Rather, the locus of responsibility lies in the word giving itself. When I promise you that p, the claim I give you constrains my future behavior. I commit myself to behaving in the future so as to make it the case that p. When I testify to you that p, the claim I give you manifests a constraint on my current behavior. I present myself as both accepting p and as having the resources to underwrite both my and your reliance on p.[3] If p fails to be true enough, I am to blame, since I invited you to rely on p's being true enough, and I implicated that I was in a position to issue such reliance. I mis-testified that p. The mis-testifier is blameworthy for having given her word in the first place, for having invited the word taker to rely on it. The proper reproach then is something like, "You shouldn't have said it if you were not sure." Mis-testifying is morally wrong because it is epistemically wrong.

Infringing a Right

It is irresponsible to invite someone to rely on your word when your word is not trustworthy. But when is that? One might think that a person's word is trustworthy whenever what she says is true enough and is untrustworthy whenever it is not. In that case, my promise is trustworthy whenever I do what I promise to and untrustworthy whenever I do not. It is not clear that we should say this though. Suppose I promised to meet you at the South Station at five o'clock, but I had no intention of keeping my promise. Or suppose that although I intended to keep my word, I was, and should have known that I was, unlikely to be able to do so. In the last ten years, the noon train from New York has almost never been on time, as it would have to be for me to be able to keep my promise. As it turned out, though, my train was early. So I encountered you in the station at five o'clock. Thomson contends that I infringed no claim of yours, since I kept my word. She takes it that the claim my promise gives you lies in the truth of p and thus is not infringed so long as p turns out to be true (1990:305–306). I am not convinced. It seems plain that you ought not to have counted on me. It was too nearly a coincidence that we met at the appointed time and place. My word was untrustworthy. Similarly, if I testify on inadequate grounds that the prehistoric cave paintings served a political purpose, even if it turns out that I am right, my word is untrustworthy. You ought not base cognitively significant inferences or actions on it.

Word Giving   151

Should we say, nevertheless, that my testimony did not infringe your claim? Even if Thomson is right about promising, I do not think that we should. To see why, we need to consider the point of each practice. Promising is future directed and action oriented. Because we in fact met at the station at five o'clock, I did what you were counting on me to do. Hence I did not cause your plans to go awry. Whether or not I ought to have given my word as I did, perhaps I infringed no claim, for I (per accidens, to be sure) kept my word. Testimony's epistemological function is to serve as a conduit of epistemic entitlement. A speaker cannot convey epistemic entitlement if she has none. The mere fact that her statement is true is not enough to epistemically entitle her to it. It could just be a lucky guess. If, purely on a hunch, I testify that the cave paintings served a political purpose, I have no evidence to back that statement. Hence I have no epistemic entitlement to convey to you. This suggests that a testifier infringes a word taker's right when she testifies to something for which she lacks sufficient grounds.

Let's look at it from the word taker's perspective. If I accept someone's testimony, I do so because I take her to be warranted in saying what she does. If it is reasonable for me to accept her testimony, it is reasonable for me to consider her warranted. In that case, I don't just accept that she means what she says. Nor do I accept merely that she considers herself warranted. Rather, to take her word for something involves accepting that she actually is warranted—that she has adequate grounds. The question then is what constitutes adequate grounds? A seemingly obvious answer is that adequate grounds consist of evidence or reasons that are sufficient to support the statements that constitute the testimony. But this is not enough. Unless there is good reason to think that the evidence or reasons are adequate, we ought not take her word. Suppose a blood test reveals the presence of antibodies that are in fact antibodies to a newly discovered virus. Skeptical worries aside, the antibodies are sufficient evidence of the virus. Dr. Norris testifies on the basis of the blood test that Zach has the virus. Absent medical consensus that the antibodies in question are the antibodies to that particular virus, Dr. Norris, although speaking the truth and having what is in fact adequate evidence, lacks warrant, for she does not satisfy the standards of the medical community. Until the connection between the antibodies and the virus is established to the satisfaction of that community, we ought not take her word.

Should we take someone's word if the evidence she relies on satisfies the standards of the relevant epistemic community, even if the evidence turns out to be misleading? Suppose Professor Cro testifies on the basis of the best available evidence—evidence that satisfies the standards of the paleoanthropological community—that the cave paintings are seventeen thousand years old. The best currently available evidence is circumstantial. There is, to be sure, an acknowledged penumbra of vagueness around the dates that paleontologists assign. But the experts are confident that seventeen thousand years old is about the right age, and they have good reason for their confidence. Suppose, though, that they are wrong. If the paintings are in fact twenty thousand years old (an age that lies well outside the penumbra of vagueness), should we consider Professor Cro epistemically blameworthy for having testified as she did? Does her testimony infringe a claim?

We can and should hold people blameworthy for testifying on the basis of insufficient evidence. If, purely on the basis of scattered anecdotal reports or an experiment run on just twelve subjects, a scientist testified that drinking green tea cures poison ivy, we would consider him epistemically remiss. But arguably the case we are considering is different, for Professor Cro had what everyone concedes was excellent evidence. We might, of course, take a hard line. You have a right to remain silent. So anything you say can be held against you. Despite the best efforts of the community of paleontologists, which were in fact quite good, Professor Cro's testimony misled scholars who took her word. Hard liners would insist that responsible testimony, like knowledge, requires truth or anyway something very close to the truth. If so, she should not have testified as she did.

If we take the hard line, testimony that turns out not to be true enough violates a right, even if, at the time of the testimony, there is no reason to doubt its adequacy and overwhelming reason to consider it adequate. Perhaps the counterexample to a highly confirmed universal generalization has not yet even arisen. Perhaps the methods required to discredit it have not yet been developed. Nevertheless, if I give you my word that p and in fact $\sim p$, I infringe your claim. Such a hard line might seem to violate the maxim "'Ought' implies 'can.'" If I genuinely could not have known or reasonably suspected that p is not true enough and/or that the evidence for p is misleading, then I was under no obligation to deny that p. Hence, it may seem, I ought not be faulted for testifying that p. But things are not so simple, for I need not have testified at all. Perhaps I could not have known that p is false. But I surely could have known—indeed, surely did know—that p might not

Word Giving 153

be true enough. I could simply have held my tongue. "'Ought' implies 'can,'" then, does not directly discredit the hard line.

We can avoid imparting falsehoods by exercising our epistemic Miranda rights. In normal epistemic contexts, testimony cannot be compelled. But withholding testimony has a price. In hoarding information, we impede joint activities and lose opportunities to advance understanding through education, collaboration, testing, and extrapolating from other people's findings. It is irresponsible to testify without adequate evidence. It may be equally irresponsible to be excessively demanding about matters of evidence. There is a familiar tension between the desire for actionable information and the requirement that the information consist entirely of truths. Reasonable levels of evidence tend to be satisfied by falsehoods as well as truths. A critical question is where the balance should be struck (see Riggs 2003). If we raise our standards enough to eliminate the falsehoods, cognitively valuable truths are excluded as well. We cannot evade this difficulty by demanding that the information be merely true enough, for a gap between being supported by available evidence and being true enough remains. The parallel to arguments that push us toward skepticism is plain (see Adler 1981). We can avoid judging falsely by refraining from judging at all. We can avoid mis-testifying by refraining from testifying at all. But refusing to accept and refusing to testify are cognitively costly. The risk of error is sometimes worth taking. Nevertheless, if the hard line is correct, I put myself morally and epistemically in jeopardy every time I testify. That gives me an incentive to increase the level of evidence I demand. To protect myself from inadvertent wrongdoing, I don't just want adequate grounds—I want grounds that I am confident are adequate. That is a more demanding standard. If it must satisfy the hard liner, it may be an unsatisfiable one. If Dr. Cro was blameworthy, despite the fact that the test needed to discredit her report had not even been invented at the time she testified, I should hardly be complacent merely because my remarks satisfy contemporary standards. The worry is that the hard line, by supplying a disincentive to testify, stifles information transfer at the frontier of inquiry.

A similar worry can be raised about promising. If my failure to keep my promise, for whatever reason, puts me morally in the wrong, I should be extremely circumspect about making promises. Before I give my word, I should be absolutely sure I can deliver. Unfortunately, I cannot be absolutely sure. Neither can anyone else. Should we stop making promises? Given the utility of the practice, that seems a high price to pay. Luckily, we don't have to pay it. Granted, we shouldn't give our word cavalierly, but obsessive caution is not required. When I make you a promise, we both recognize that I *might* not be able to keep it. Unforeseen circumstances might interfere. Even if I am scrupulous about my moral character, that recognition should not prevent me from giving my word, for part of the institution of promising is that there are forgivable lapses and acceptable excuses. If I fail to keep my promise to meet you to go comparison shopping for grass seed, I infringe the claim I gave you. But if the reason for my absence was that I was negotiating with a deranged student who was holding the dean hostage, my failure to keep my word is excusable. Perhaps I owe you an explanation, but I owe you no apology, since we agree, and recognize that we agree, that that sort of demand on one's time takes precedence.

We might want to say the same about testimony. Although testimony that fails to be true enough infringes the word taker's claim, there are forgivable lapses and acceptable excuses. You exonerate me for promise breaking, saying, "You couldn't have known." My lapse is excusable, for there was no way I could have foreseen the hostage situation that prevented me from keeping my word. The message is this: Had you known that p when you said what you did, you would have been seriously remiss. But since you couldn't have known, you are morally and epistemically off the hook. We might want to make the same sort of move in the case of false but well-grounded testimony. Had Professor Cro known that the cave paintings were twenty thousand years old or had considerably more accurate dating methods been available and accessible, she would have been seriously remiss when she testified that they were painted seventeen thousand years ago. But since she couldn't have known—since the relevant test will not be developed for another fifty years— her error is excusable. We can then retain the hard-line requirement that the content of testimony must be true enough, while weakening the disincentive to testify by conceding that some testimony that fails to be true enough is excusable.

One might wonder whether concern with (something in the neighborhood of) truth is an idle wheel. In deciding whether it is reasonable to give or accept testimony that p, we consider whether the assertion that p is well founded—that is, whether it satisfies the requirements of the relevant community of inquiry. Even though we recognize that well foundedness provides no assurance that it is true enough, we don't and can't go on to ask the

Word Giving

further question: Besides being well founded, is *p* also true enough? Our best hope of discovering whether *p* is true enough lies in discovering whether *p* is well enough founded. Current standards of acceptability are the best standards we have for deciding that. It makes no sense then to construe the truth requirement as an additional factor that figures in the decision whether to give or to accept testimony that *p*. Nevertheless, it does not follow that concerns about whether a contention is true enough are idle. They play a different role. Testimony is responsibly proffered and accepted when it satisfies the current standards of the relevant epistemically reputable community of inquiry. Subsequently, new evidence, improved techniques, or refined standards may lead us to conclude that previously accepted testimony is not true enough. If its failing to be true enough is a sufficient reason to reject it as error, we have the resources to construe revisions in the considerations we accept, the methods we use, and the standards we hold ourselves to as improvements rather than mere changes in our understanding. If the best we can say is that *p* satisfied the standards accepted at one time but not those accepted at a later time, we do not, for in that case changes in what it is reasonable to accept or to testify are like changes in fashion. Sometimes one standard or skirt length is in style; sometimes, another. A truth requirement is not the only requirement that could play this role. Nor is it clearly the best choice. An acceptability requirement might be preferable. But some such requirement is needed to distinguish advancing understanding from merely changing intellectual fashions.

Uptake

Word giving, according to Thomson, requires uptake. The invitee, she says, needs to receive and accept the invitation to rely on the truth (or, I would say, the acceptability) of *p*. But, it seems, we are inundated with testimony we have no use for. Textbooks, news reports, lectures, and gossip supply vast amounts of seemingly useless information. Does this discredit Thomson's account? To decide, we need to consider what accepting an invitation involves. To accept my invitation to dinner on Sunday at seven o'clock requires appearing for dinner on the appointed day at roughly the appointed time. To accept my invitation to call on me if you need help is different. You accept my invitation if you henceforth consider yourself free to call—if, that is, you adjust your attitudes so that requesting my help is now a live option for you. You may turn out not to need my help. But even if no call is made, the invitation is accepted. Testifiers issue invitations of both kinds. My testimony may provide you with the specific information you need for a particular purpose. I inform you that in the 1760s, Hume was a diplomat in Paris. Relying on my expertise, you incorporate that information into your book on the history of Scottish thought. But not all information transfer is on a need-to-know basis. I make the same statement in an introductory philosophy lecture. I invite my students to accept it, just as I invited you. Most of them will do nothing with it. They have no use for it. In my lecture, I issue an open invitation. I invite my students to rely on my contention when and if they need to. If they are prepared to do so, they accept my invitation. Both of these sorts of reliance fit Thomson's model easily. The argument that we receive vast amounts of useless information does not discredit her analysis.

What should we say about proffered testimony that is flatly disbelieved? The invitation to rely is issued, received, and rejected. Should we say that such testimony is abortive? If so, there is no word giving without word taking. That seems wrong. The suspect's mother asserts under oath that he was home watching television at the time the crime was committed. No one believes her. Still, it seems, she testifies that he was home. She couldn't be charged with perjury if her statements did not amount to testimony. But simply to jettison the uptake requirement also seems wrong. If my students sleep through the lecture where I say that Hume was a diplomat, or I make that statement in a language they don't understand, we would be reluctant to say that I gave them my word that Hume was a diplomat. They can't take my word for it, since they have no idea what my word is. I recommend then that the uptake requirement be modified. Testimony is abortive, I suggest, unless the invitation is received. But the invitation need not be accepted. Receiving an invitation to rely on the acceptability of a statement is not just having one's sense organs stimulated. It requires understanding the statement's content and force and recognizing that one has been invited to rely on its acceptability. When these conditions are met, the invitation has been received. There is uptake whether or not the invitation is accepted.

Cooperation

If I testify that p, I warrant that p is acceptable. But if I testify that the cave paintings are seventeen thousand years old, I do not commit myself to the

Word Giving

157

truth of the sentence "The cave paintings are seventeen thousand years old."
I would be astounded if they were exactly seventeen thousand years old. I
would consider myself, and be considered by others, to be right if I was off
by no more than a few hundred years. Indeed, without new evidence, I am
apt to utter the very same sentence in my lectures year after year. If I thought
the paintings were exactly seventeen thousand years old this year, I should
update my notes and say that they are 17,001 years old next year. Evidently,
I use a seemingly precise sentence to convey a considerably vaguer message.
It is the acceptability of the vague message, not that of the precise sentence,
that my testimony commits me to. There is nothing disingenuous about this.
I am not pretending to provide more precision than I do. It is tacitly acknowledged on all sides that the age I ascribe has a fairly generous penumbra of
vagueness. If the actual age of the paintings falls within the penumbra, my
testimony is acceptable.

How then is it that the message conveyed is the message received? What
prevents my audience from concluding that my testimony reports the exact
date the paintings were produced or from ascribing to it a significantly different penumbra of vagueness? If all parties to an exchange belong to the same
epistemic community, there is no mystery. In that case, everyone imposes
the same constraints on the interpretation of my words. But why should we
think this? If the assumptions have not been expressly agreed to, why should
we think that they are shared? Background assumptions plainly vary from
one linguistic context to the next. Moreover, they are continually revised and
updated as discourse proceeds. But they are neither random nor idiosyncratic.

Grice's account explains why. His basic insight is that communication is
fundamentally interpersonal. This does not sound particularly momentous,
but he shows that it is a deep and deeply important point. An informant is
not just a spouter of information, nor is a receiver an empty vessel into which
data are poured. Because every interchange involves background assumptions, speaker and hearer must understand each other. This is not just a matter
of grasping the words that comprise an utterance or inscription. It involves
appreciating why, to what end, and against what background those particular words are uttered or inscribed. To understand an utterance requires understanding its utterer, for communication is a matter of mutual attunement.
Epistemic empathy is required. This is why, Grice contends, communication
depends on cooperation. Informative exchanges are, he maintains, governed
by the Cooperative Principle: "Make your conversational contribution such

as is required at the stage at which it occurs, by the accepted purpose or direction of the talk exchange in which you are engaged" (1989:26). This requires that the contribution be informative, relevant, true enough, and supported by evidence.

Although the Cooperative Principle and the associated maxims are cast as instructions to the speaker, they supply rules for the hearer as well. Ceteris paribus, in order to interpret an informative utterance or inscription correctly, we construe it as one that satisfies or at least purports to satisfy Gricean maxims. In a communicative exchange, each party not only conforms her contributions to the maxims (or flagrantly flouts them) she also takes it that the other parties are doing so. Interpreting then is not a matter of rote application of the homophonic rule or of some regimented principle of interlinguistic translation. It involves consideration of what interpretation of the speaker's remarks would be one that the speaker could have, or at least believe herself to have, adequate evidence for, what interpretation would yield a statement that the speaker would consider informative, relevant, and apt. You don't take me to have testified that the cave paintings are *exactly* seventeen thousand years old because you don't think it remotely plausible that I have evidence that could support such a precise statement, nor do you think that such precision is required or even desirable in the context in which we are speaking. You take me to have testified that the paintings are in the neighborhood of seventeen thousand years old, since that is an informative, contextually relevant contention that you think I could have adequate evidence for. You also deploy the maxims in assigning the neighborhood a size. What all sides concede goes without saying, for if all parties agree that p, "p" is uninformative. Therefore, you take me to be saying something more revealing than what everyone in the audience already knows anyway. Considerations of relevance provide further constraints. If the discussion requires that the date be specified within five hundred years, I am uncooperative if my remark is not that specific. Since you take me to be cooperative, you therefore interpret my remark as saying that the paintings are within five hundred years of being seventeen thousand years old. If we only need a date within five hundred years of the right one, it would be uncooperative of me to be much more precise than that. So, you have reason to refrain from taking my statement to be overly precise.

Gricean considerations show how complex and context sensitive uptake is. To interpret a speaker's testimony properly involves an awareness of the course and point of the discussion, as well as an appreciation both of what

Word Giving

has already been established and of what goes without saying. The speaker purports to be satisfying the evidence maxim. So we need to construe her as saying something she has or takes herself to have or purports to have adequate evidence for. To do that, we need to be sensitive to the relevant epistemic norms. We need, that is, to understand what sort of evidence and how much evidence is required. To decide among the available interpretations of a speaker's words requires recognizing which of them she can purport to have adequate evidence for, hence what evidence she might have and what evidence would be adequate. Evidential standards vary. A measurement that would be acceptable in the kitchen is apt to be too rough to accept in the lab; the precision expected in the lab is too fussy for the kitchen. Finally, we need interpersonal awareness. It is not enough to know what has actually transpired in the course of the discussion and what is actually required by way of evidence. We also need to understand what each party takes to have transpired and what each takes to be required.

To understand someone's testimony is to construe it as a statement of fact (or a collection of statements of fact) that the testifier purports to accept and to have adequate grounds for. To be sure, people sometimes testify without adequate grounds, being either misleading or misled about the strength of their evidence. In taking someone's word, we assume that she is neither. This might be doubted. Suppose Pat, for no good reason, says that p. Although Max realizes that Pat has no justification for her remark, he has very good reasons for believing that p—reasons he never brought to bear on the issue prior to hearing Pat's totally unfounded utterance. He is now justified in accepting p and came to be justified via Pat's statement. Still, Max does not take Pat's word that p. Pat's statement was a catalyst but conveyed no epistemic entitlement. He did not accept Pat's invitation to rely but took the occasion to marshal his own evidence. A harder case is this: Suppose Sasha testifies, on relatively weak grounds, that q. Sasha's grounds are inadequate. But they're not *nothing*. They afford some reason to believe that q. Jim has additional grounds that are also insufficient if taken alone. But combined with Sasha's grounds, they yield sufficient evidence for q. Jim relies partly, but not wholly, on Sasha's testimony. I suggest that the strength of his reliance on Sasha's testimony is determined by the strength of the backing he takes Sasha's testimony to have. Word taking then can be a matter of degree. We may partly rely on the word of someone whose evidence we consider weak and augment that reliance with reasons of our own.

We are justified in taking someone's word only to the extent that we are justified in thinking her reasons are adequate. But we can take a speaker's word and be justified in doing so without knowing what her reasons are. In some cases, a speaker's behavior and/or the mundane nature of the information might afford ample evidence that she is satisfying the Cooperative Principle, hence that she has what she takes to be adequate evidence. If the topic is mundane enough, her taking herself to have adequate evidence is reason enough to think that she does. Unhesitatingly reporting one's zip code is a case of this kind, since normally a speaker knows her zip code and typically lacks any reason lie about it. In cases where evidence of cooperation is not enough, we may know the particular speaker to be morally and epistemically trustworthy. Then even though we lack access to her reasons, we know that she would not testify if they were inadequate. In yet other cases, testimony may be given in a context where there are sufficient institutional safeguards to block epistemically irresponsible testimony. The fact that the experts in the field raise no objection indicates that the evidence, whatever it is, satisfies the relevant standards. If the field is epistemically estimable, institutional safeguards are safeguards enough.

Carrying the Load

Testimony then conveys more than the information that comprises its content. It also implicates that the facts it reports have been established to the satisfaction of the relevant community of inquiry and that the testifier is in a position to epistemically entitle her audience to accept them. That being so, a speaker testifies responsibly only if she is in a position to shoulder the epistemic burden for everything her testimony conveys. It might seem that this does not add to the load. Perhaps a speaker is epistemically entitled to convey anything she is epistemically entitled to accept and epistemically entitled to accept anything that satisfies the standards of the relevant community of inquiry. If so, the brute fact that she has adequate reasons suffices. She need not be aware that her reasons are adequate. She need not even be aware of what her reasons are.

This is in line with epistemological theories that hold that a subject can be fully warranted in believing that p without being aware of what supplies the warrant. Such theories provide an attractive account of perceptual warrant. Seeing a rabbit twenty feet away in the center of his visual field wholly justifies a subject with good eyesight in believing that there is a rabbit in

Word Giving

161

front of him. He need not have the conceptual resources to appreciate that his perception supplies him with grounds, much less know anything about the perceptual mechanisms that make seeing reliable. At least in some cases then, there is reason to think that it is the having of grounds not the awareness that one has grounds that is required for warrant. But even if this is so, and even if it holds for warranted belief generally (which I doubt), nothing directly follows about what is required to convey warrant.

Being in a position to convey warrant requires more than merely being warranted. A subject who has scattered evidence that warrants her acceptance that p but has never put that evidence together does not realize that she is warranted in accepting that p. She is in no position to give her word that p, since she is not prepared to shoulder the epistemic burden that comes with testifying that p. A subject whose evidence in fact warrants q might fail to realize that her acceptance that q is warranted because she thinks that stronger evidence is required. Again, it seems, she is unable to shoulder the epistemic burden, since she considers her grounds inadequate. These examples suggest that in order to testify responsibly, an epistemic agent must be justified not only in accepting p but also in accepting that she is justified in accepting p.

This sets an additional demand but not an unsatisfiable one. It does not require ever more evidence for p. Rather, it requires reason to think that one's evidence or reasons for p are adequate. It therefore introduces second-order considerations about the adequacy of reasons. If Jim is to be justified in accepting that he is justified in accepting that p, he needs to appreciate the basis of his acceptance. This requires critical self-awareness. He needs self-awareness because he must be cognizant of the acceptances and perceptual states that supply his grounds. The self-awareness must be critical, for he must recognize that the considerations he adduces qualify as reasons to accept that p. The fox is warranted in believing that there is a rabbit in front of him but is not justified in believing that his belief is warranted, for he has no idea why he trusts his senses or whether it is reasonable to do so. Jim also needs some awareness of the relevant epistemic standards. He has to know what sort of evidence and how much evidence is required in a context like this to support a commitment like p. He needs, moreover, to credit those standards. He must consider them reasonable or at least not unreasonable. If he considered the accepted standards of evidence to be epistemically shoddy, he would have no reason to take their satisfaction to confer epistemic entitlement. Recognizing that one's reasons satisfy the standards of the contemporary astrological community does not inspire confidence in the contention they are supposed to support. Finally, he needs to recognize that the grounds he has satisfy the relevant epistemic standards—standards that he reflectively endorses.

This is fine, one might think, if we are talking about the first link in the chain of epistemic entitlers—the person who actually marshaled the evidence. If a subject is attuned to the standards of the relevant community of inquiry, recognizes that they are reasonable standards, reflectively endorses those standards, and realizes that her evidence satisfies them, she justifiably accepts that she is justified in accepting and in testifying that p. Often this is too much to ask. As an intermediate link in the chain, Mike is justified in holding that p and undertakes to pass the information along. He read it in the newspaper, heard it in a lecture, learned it in school. But he is in no position to supply the backing for it. Nor does he have the expertise to recognize or endorse the standards of the community that underwrites his belief. Still, one wants to say that as an informed layman, he can testify responsibly that the political situation in Rwanda is unstable, that electrons have negative charge, that Hume was a diplomat. An informed layman is not a gullible stooge. He recognizes that he has good reason to think that the experts his judgment depends on are trustworthy. The source he relies on to back up his assertion is not just a trustworthy source it is also a source he considers trustworthy and has good reason to consider trustworthy. He then has reason to think that his information belongs to a trustworthy communicative chain. Even intermediate links in the chain of epistemic entitlers then satisfy the demands of critical self-awareness.

Testimony turns out to be more complex than the idea of information transfer might initially suggest. Testifying that p is not merely asserting or professing that p. Nor, of course, is testifying that p the same as testifying that one is warranted in testifying that p. But it would be unreasonable for you to take my word that p if I was not warranted in professing that p. When I testify to you that p then, I do not merely impart the information that p is the case. I also give you reason to believe that p is warranted and that I am warranted in professing that p. My testimony gives you moral and epistemic claims against me. If p is not true enough (and no exonerating conditions obtain), then in testifying that p, I both impart inaccurate information and do you a moral wrong. I mislead you about p's epistemic standing by assuring you that it is epistemically safe to rely on p when in fact it is not. That leaves you vulnerable. So the ground for the moral wrong is an epistemic wrong. In the realm of rights, epistemology and ethics overlap.

7 Word Taking

Introduction

The testimony we have considered concerns information transfer between differently situated but otherwise epistemically equal parties. Sally knows that the subway goes to Alewife; Harry does not. But otherwise they are epistemic peers. In some cases of information transfer, however, the parties are not peers. One has a deeper, broader, more textured understanding of an issue and the viable ways of approaching it than the other. Physicians, plumbers, engineers, and dietitians are experts in their respective fields. They do not just have more information than the rest of us; they are, we think, more proficient. They know how to approach certain issues and think about them in ways that we do not and often cannot. They may be, we suspect, in relevant respects just smarter than we are. This need not involve possessing credentials or degrees. They may have come by their understanding simply through experience that led them to scrupulously examine, attend to, and investigate matters that we never did. With respect to particular issues then they are our epistemic superiors. What is the nature of that superiority? How should we respond on finding that someone with a considerable epistemic advantage believes that p?

Often such a person is described as an "epistemic authority" vis-à-vis p. But we should approach such a description with caution for, I will argue, authority is different from expertise. It might seem that this is simply a terminological matter. Perhaps it makes no difference whether we construe epistemic superiors as authorities or experts. I am not terribly concerned about the language we use. But there is a distinction that matters epistemologically.

Authority

To have authority is to possess legitimate power to do something, where that power derives from, and is circumscribed by, a role in a social and/or political practice. In suitable circumstances, a judge has the authority to impose a fine; a referee has the authority to call a foul; an officiant has the authority conduct a wedding; a professor has the authority to give grades. An authority is someone who is vested with specific, legitimate powers by and within a particular practice. The powers carry permissions, prohibitions, rights, and obligations that others—even other participants in the practice—lack. A judge can impose a fine; a bailiff cannot. A referee can call a foul; a player cannot. Although we hope that those in authority display relevant competences, competence is not required. A referee can make a bad call, yet the foul stands. He can make consistently bad calls, but until his authority is rescinded, he retains the right and the responsibility to call fouls. A professor can give unreasonably harsh grades, yet so long as they fall within the broad parameters the university sets, the grades stand. In some practices, under some conditions, it is possible to appeal an authority's verdict to a higher authority. But at the end of the chain of appeals, an authority settles the matter.

Nevertheless, an authority's action is open to criticism. Because it was the act of someone vested with the legitimate power to act, the verdict stands. Still, students justifiably complain about unduly harsh grades; sports fans justifiably disparage bad calls. As participants in the practice, they are obliged to accede to the verdict, but they need not—and often do not—go quietly. Critics are often knowledgeable. They understand the practice and the way it is designed to promote certain goods. They understand why ill-advised or incompetent exercises of authority impede the promotion of those goods. Often the calls that they deem bad actually are bad. Exercises of authority are thus subject to epistemic assessment.

A related concept is authorization. To be authorized is to be vested with authority. Spokesmen are authorized to speak for others. Agents are authorized to act on behalf of their clients. The power to speak or act on their behalf is conferred by the authorizers—the individuals or organizations the agent acts for. In advising Americans to get vaccinated, the head of the CDC speaks for the Centers for Disease Control and Prevention; in agreeing to the terms of the book contract, the literary agent acts on behalf of his client. Again, we hope that the authorized parties are competent, but that is not

Word Taking

required. So long as the agent acts within the limits of the authority granted to him, the authorizer is bound by the commitments made on his behalf. Here too it is possible to criticize. The spokesman should not have said that; the agent should not have done this. Still, the authorizer is answerable for what its agents said and did in his name.

The conception of authority I have been discussing is independent of epistemology. Practices define roles and confer certain powers on those who play those roles. They determine under what circumstances one person can be authorized to speak or act for others—to call a foul, impose a fine, or conduct a marriage ceremony. Those who play those roles need not have any special talents or abilities. They may play the roles well or badly. It makes no difference. If the myopic ref calls a foul, it is a foul. If the incompetent agent signs the contract, her client is still bound by its provisions. The critical issue is that authority is vested in someone to do something.

We sometimes speak of epistemic or intellectual authorities. Roger Penrose, we say, is an authority on black holes; Helen Vendler, an authority on Shakespeare's sonnets. This locution is ill-advised. There is no reason to think that Penrose and Vendler have professional rights to do anything that other physicists and critics are barred from doing. There is no reason to think that they speak or act on behalf of anyone but themselves. They are just more intelligent, more knowledgeable, more acute, more creative than the rest of us. To call them authorities is to suggest that they have legitimate powers to do things that others are barred from doing. That is false. Moreover, their claims can be, and often are, challenged. Some of the challenges are well founded. Then the acceptability of their claims is undermined. I suggest that they are experts, not authorities.

Expertise

Expertise consists in epistemic achievement. An expert is epistemically considerably more adept than those who rely on her. She has greater understanding of a topic, greater facility in solving problems in the area, greater adeptness in finding, filtering, and exploiting relevant information than those who depend on her. Although expertise is often signaled by credentials, having suitable credentials is neither necessary nor sufficient for qualifying as an expert. Nor is expertise an absolute measure of epistemic achievement. It is a comparative state, deriving from there being a sizable gap between levels of epistemic achievement of different people. Some have greater understanding, know-how, or knowledge than others. As a result, a person's understanding of a given subject matter may qualify her as an expert for one group, an epistemic peer for another, and an epistemic inferior for a third.

Expertise is specialized. An expert in one area—plumbing or medicine for example—is a novice in others—maybe astronomy or poetry. It might seem then that to qualify as an expert in an area, an agent must be considerably more adept than most people. This is not quite right. Plenty of first-time parents justifiably treat more experienced parents as experts about childcare, seeking advice about matters like how to soothe a teething baby and information about when he is likely to sleep through the night. Although such information is widespread, the epistemic distance between the experienced and the inexperienced is large enough to qualify as a gap in expertise. Rather than saying that experts need to be more accomplished than most, we should say that they are considerably more epistemically accomplished than those who seek to draw on their epistemic accomplishments. Being rightly treated as an expert is, evidently, more basic than being an expert tout court. What is crucial is that, unlike experts, novices cannot—or are ill-equipped to—find out or figure out the relevant facts for themselves.

Even in areas where knowledge is sparse—for example, about the behavior, distribution, and constitution of dark matter—there is a gap between novices and experts. Pretty much all that anyone *knows* about dark matter is that there must be a lot of it to account for the behavior of the detectable matter in the universe. Still there are experts—those who know how to investigate the issue, how to entertain and assess hypotheses, and so on. The threshold for expertise is thus fluid, varying with background information, the seeker of information, and the epistemic accomplishments of the provider of that information.

Expertise is epistemically valuable because experts can provide novices with trustworthy resources for expanding their epistemic range. They can provide information or solve problems that others cannot easily solve for themselves. Goldman holds that this is a matter of experts having a higher proportion of true beliefs and a lower proportion of false ones than the public at large (2018:4). I'm not convinced. Experts about dark matter probably have a larger proportion of false beliefs about the topic than the public at large because the public at large has few if any beliefs about the topic and those they have are relatively generic beliefs which are more likely than the expert's more refined

Word Taking

views to be true. Still, I agree that experts are epistemically more accomplished than novices, and they can share (some of) their accomplishments with novices via the transfer of information and relevant skills.

I said earlier that the difference between an epistemic authority and an expert may be merely terminological. The term "authority" then would be ambiguous—in the one case, a matter of powers and obligations that derive from a role in a social practice; in the other, a matter of epistemic accomplishment. This would accord with usage. But whether or not we regiment our terminology, we should take pains not to confuse the social with the epistemic conception.

Epistemic Authority

Zagzebski (2012) holds that the concept of authority is univocal and that it applies to those with expertise. If she is right then, because Alice is an epistemic authority with respect to a particular proposition p, Alice's thinking that p constitutes p's being justified for those who lack the relevant authority. Just as the referee's calling a foul makes it a foul, the epistemic authority's considering a claim justified makes it justified. To recognize someone as an epistemic authority on this view is to vest him with the power to settle epistemic issues. Zagzebski endorses the *Preemption Thesis for epistemic authority*:

> The fact that the authority has a belief p is a reason for me to believe p that replaces my other reasons relevant to believe p and is not simply added to them. (2012:107)[1]

Several features of this thesis are worthy of attention. First, it is a replacement thesis. That an epistemic authority believes that p replaces rather than merely augments my other relevant reasons. It does so preemptively; it blocks any consideration of other reasons. Second, it is the mere fact that an authority believes that p, not my knowledge or awareness that he does, that underwrites replacement. Third, she says that the authority's belief is *a* reason to believe p, not *the* reason (see Jäger 2016:167–168). But if the fact that A believes that p replaces any reasons that I might have to believe that p or to believe that ~p, A's belief is, as she treats it, a conclusive reason.

Zagzebski holds that epistemic authorities can command belief—that is, rightly require that their claims be believed (2012:101). I am not convinced that we can believe at will. So I am not convinced that we can believe on command. She maintains that speakers command belief when they say "Believe me: *p*." The fact that we issue such commands, she thinks, is evidence that they can be obeyed. She is, of course, right that we sometimes use such locutions. But it is not obvious that the speech act we thereby perform is issuing a command. Rather, I suggest, "Believe me" functions to advise, recommend, or even implore. That makes acceptance optional. The phrase, when sincerely uttered, indicates that the utterer is strongly committed to the correctness of *p* and thinks it desirable that the hearer accept it. The most that follows is that the speaker holds that the hearer stands to make an error if she does not agree. It does not follow that the hearer is obliged to agree or even that she should agree. Every spring, Boston sports fans are apt to say, "Believe me: The Red Sox will win this year," when no one knows who will win (and the team's prospects are bleak). "Believe me" emphasizes that the utterer stands behind the claim, but it does not by itself indicate that he ought to stand behind it, much less that others are epistemically obliged to acquiesce. Even if Alfred is epistemically more accomplished than Sue in a given area and can appreciate that Sue will make a big mistake if she does not believe what Alfred does, it does not follow that Alfred can command agreement.

Zagzebski denies that believing on the basis of authority is believing on the basis of testimony. The mere fact that an epistemic superior believes that *p* gives a novice conclusive reason to believe that *p*. The novice need not entertain *p* (that is, consider whether *p* is true or is credible), nor does he have to grasp the content of *p*. Even if he does not understand what *p* says, the fact that an epistemic superior believes it constitutes a conclusive reason for him to agree. I should thus believe everything written in the current issue of *Science*, inasmuch as the authors are all my epistemic superiors in scientific matters, and I should do so, even without scanning the table of contents. More alarmingly, I should believe whatever an epistemic superior believes, even if it is not vetted via peer review—indeed, even if it is not communicated.

Suppose there is a scientific institute in a remote valley in the Himalayas where epistemically superior investigators pursue their inquiries. One spends his life investigating an obscure topic—perhaps the gastric juices in the common housefly. He comes to believe that the gastric juices have a particular chemical composition. I have no epistemic access to his findings. Still, Zagzebski maintains, I should believe what he does. I have never given the issue a moment's thought. So I have no antecedent views on the subject. If I follow his lead, I come to have an additional well-founded belief. Of course, I have

Word Taking

no reason to think this is the case. Nevertheless, I should believe what he does. Suppose another scientist studies something I have casually looked into so I have views on the subject. Perhaps the issue is whether the severity of winters has any bearing on the date when the first crocus blooms. I made no systematic investigation. Rather, I drew a rough-and-ready correlation based on crocuses in my yard. Again, I have no access to his research. Nevertheless, I should, Zagzebski maintains, scrap my view and my evidence for it and substitute his belief. What Zagzebski advocates is blind, ignorant acquiescence. If reliabilism is correct (which she thinks it is), then such blind, ignorant acquiescence often yields beliefs that constitute blind, ignorant knowledge.

Reliabilism is a form of consequentialism. The good it aims to realize is believing as many truths as possible and disbelieving as many falsehoods as possible. Reasons and evidence are merely instrumentally valuable. They should be set aside if a better instrument becomes available. Blind allegiance to the views of an epistemic authority is, Zagzebski maintains, a better instrument than relying on my own evidence and reasoning powers. But if I do not know or reasonably believe that an epistemic superior believes that p, I have no basis for believing that p. All the reasons I have access to, we assume, do not support that p. Granted, my evidence is limited and may be skewed. Still it seems better to believe (or disbelieve or suspend judgment) on the basis of that evidence than on the basis of something that is not within my ken.

If I do not understand the content of p, then it is not at all clear what believing it comes to. Consider, for example, "the Na+NaLi system was shown to have only ~4% loss probability in a fully spin-polarized state" (Son et al. 2022:1006). I have no idea what that means. Dr. Son is clearly my epistemic superior with respect to quantum chemistry. So, Zagzebski says, I should believe what he does in this area. Typically, to believe that p is to hold that p is true. It is to hold, roughly, that the world is such that p. Someone who believes that p is, within limits, willing and able to use it in nontrivial inferences and as a basis for action. But if I am completely clueless about what the content of p claims, then I am incapable of nontrivial inferences or actions on the basis of it. "Ought" implies "can." It is not possible for me to believe something that is unintelligible to me. So it cannot be the case that I ought to believe it.

What of the case when I have some grasp of the content and recognize that an epistemic superior thinks otherwise? In one case, I understand (up to a point) what p claims, but have no antecedent views about whether it is so. Even here it is not obvious that I should blindly adopt the opinion of an epistemic superior. If I recognize that she has reasons and I have none, and the topic is not too important, maybe I would do well to adopt her view. But even where I have no antecedent opinion, serious cases are more problematic. Suppose I have no medical training and my oncologist tells me that my cancer is small-cell pulmonary carcinoma and that chemotherapy is the treatment with the best prognosis. In such cases, patients are standardly advised to get a second opinion. That my oncologist knows considerably more than I do is not a good enough reason for me to cede my epistemic authority. Before I embark on a risky course of treatment, it behooves me to find out if this is the consensus opinion among experts.

Suppose I am not a complete novice in the area. I'm not an oncologist, but I have a medical degree. Although she says that p (perhaps that small-cell pulmonary carcinoma is best treated with chemotherapy), I believe that $\sim p$ (perhaps thinking that radiation is more effective). Even then, Zagzebski maintains, my epistemic superior's opinion preempts my own. I should set my opinion and my reasons for it aside. She says that I can still hold my reasons but that they should not figure in what I believe (2012:113). I should, in effect, take them offline. What is this supposed to mean? I can easily see how I might come to think that my reasons are irrelevant or inconclusive. And I can easily see why this might be what I should come to think. But preemption seems to involve something different. Although r is a reason that supports p, and I can continue to think it is, I should ignore that fact and think that p only for a completely different reason.

As Katherine Dormandy (2018) argues, preemption seems particularly odd when I happen to agree with my epistemic superior. Suppose I believe that p and have what I consider to be good reasons to believe that p. On learning that an epistemic superior agrees with me, it would seem that I now have even better reason to believe that p. But if my reasons are preempted, then they have no effect on what I should believe. Rather than having better reasons to believe that p, I now just have a different reason. The fact that p was what I already thought, and had good reasons to think, is irrelevant. Epistemically I am in the same position as a novice who antecedently believed that $\sim p$ or who had no views about the matter.

In supporting her position, Zagzebski turns to Joseph Raz. Although her preemption principle applies even in cases where I lack epistemic access to

Word Taking 171

an epistemic superior's beliefs, Raz's principles assume that I have access. So they are more limited in scope and more plausible than the Preemption Thesis. For our purposes, Raz's first principle is the important one:

Justification Thesis for the Authority of Belief (JAB 1) The authority of another person's belief for me is justified by my conscientious judgment that I am more likely to form a true belief and avoid a false belief if I believe what the authority believes than if I try to figure out what to believe by myself. (2012:110)

My epistemic superior about a subject is considerably more likely to harbor true beliefs and eschew false beliefs about it than I am. So if I want to arrive at the truth, I should adopt his beliefs. I will suggest that even with this end in view, preemption is less tenable than it seems.

The unit of analysis is important here. Zagzebski and Raz take it that we calculate the probability of candidates for belief one by one. And, I suggest, they implicitly assume that time is short. If I need to decide on this very proposition immediately, the preemption principle may make sense. Here is such a case:

Felix, a recent, mediocre graduate of the military academy, finds a ticking time bomb in downtown Chicago. It is set to go off in four minutes. The bomb is connected to the timer with a red wire and a blue wire. Cutting the correct wire will defuse the bomb; if the wrong one is cut, the bomb will immediately explode. Felix, having had some instruction in bomb defusing, thinks that the red wire should be cut. But Lissa, his epistemic superior in the matter, is also on the scene. She believes that they should cut the blue wire.

Because it is a life-and-death situation, the only downstream consequence worth considering is whether those in the vicinity will survive. Because the bomb is about to explode, there is no opportunity to consult further. It seems that the best strategy for Felix is to believe Lissa and act accordingly. She is, after all, his epistemic superior. But notice that this case is so highly artificial that it is practically a cartoon (see Elga 2007:482–483). To say that we should blindly follow an epistemic superior, silence our own views, and hope for the best in such a case gives us no insight into what we should do in more standard cases. Recall the advice about getting a second opinion. If there is time to look further, Felix's simply replacing his reasons with Lissa's is unduly risky. He would do better to find out what others who are her equals or superiors think and maybe garner first-order information on how to defuse a bomb.

To believe that p merely because someone else believes it deprives me of information about why p, about how p, and about how my belief that p

relates to other things I might believe. Once the bomb is successfully defused, Felix should go back and consider what his reasons were and where they went wrong if they did. What related matters might he be wrong about if he was wrong about this? In order to ask such questions, he must *not* set his reasons aside. Rather he should examine them and examine the weights he attaches to them. This requires adopting a more holistic perspective. Rather than assessing candidates for belief one by one, we should survey the tapestry of commitments that they are woven into. Nothing follows about whether Felix should retain his original belief. He should consider the content of his belief, the reasons for it, the content of the replacement Lissa provides, and how adopting Lissa's opinion will affect other things he has reason to believe. He might find that although his reasons genuinely supported his conclusion, they were too weak. He might find that some of them were flawed. He might find that focusing on the color of the wires was completely irrelevant, and the factor that Lissa focused on and he overlooked was what the various wires were connected to, and so on. Preemption deprives us of epistemic resources.

Zagzebski argues for preemption on the grounds that the epistemic superior's opinions are more likely to be true than the novice's. Even so, the superior's opinion is fallible. What if it should turn out to be false or unfounded? Presumably the novice should come to believe that $\sim p$ on learning that his superior's belief that p was false. But what should he do if he learns that the superior's belief that p was unfounded—that is, if he learns that his superior's reasons are inadequate? If, prior to encountering his superior's belief, the novice believed that $\sim p$, then at that time he thought that $\sim p$ had a probability of greater than 0.50. On replacing his belief with the superior's, he had to drastically lower his assessment of the probability that $\sim p$. He therefore had to repudiate some of his reasons, downgrade the weight he attached to them, or revise his standards for how much evidence is required to support such a claim. When he finds out that her belief was unfounded, can he or should he simply resurrect his earlier reasons, the weights he attached to them, and his previous standards of evidence? However this is decided, the conviction that replacement is simply a matter of setting one's belief aside is false. Minimally, the replacer had to repudiate the belief that his reasons are sufficient to believe that p.

Even if one is a reliabilist, the conviction that candidates should be vetted one at a time is problematic. Replacing my belief that $\sim p$ with my superior's belief that p is epistemically costly. Other beliefs lose support as well. A holist would argue that rather than simply considering whether the belief that p is more likely to be true than the belief that $\sim p$, it is better to assess the truth likelihood of the constellation of commitments that will be affected by changing one's mind about p.

A final worry is that the Preemption Thesis is vulnerable to self-refutation. Epistemology is subject to its own strictures. So suppose an epistemic superior in the field of epistemology believes that the Preemption Thesis is false. Then, because he is my superior, if I accept the Preemption Thesis, I should reject it. But if I reject it, I am violating its strictures.[2] The fate of the thesis seems to hang on the hope that no epistemic superiors are internalists.

Epistemic Expertise

I said earlier that expertise is always at least in part epistemic. An expert tennis player knows how to play the game. This may be embodied knowledge that she cannot articulate. Still, it is obvious that she knows how to play. But we have been focusing on propositional matters. So I set inarticulate knowledge aside. Why is construing epistemic superiors as experts rather than as authorities preferable?

Dormandy (2018) maintains that rather than preemption, an epistemic superior's belief that is contrary to my own should be considered a rebutting defeater.[3] She assumes that the epistemic superior has considered all the evidence I have for my belief and concluded that it is wanting. But from the fact my epistemic superior and I seek to decide whether the same proposition is true, it does not follow that we draw on the same evidence. If we do not, then without examining the evidence, we cannot tell whether it constitutes a rebutting defeater or an undercutting defeater, assuming it is a defeater at all. Suppose my epistemic superior concludes on the basis of fingerprint evidence that the butler stole the spoons. I believe otherwise on the basis of my assessment of the butler's character. My superior, a forensic scientist, has better evidence than I do. So ceteris paribus, I should believe as he does. But he knows nothing of, and does not consider, the butler's character. That being so, his belief provides an undercutting defeater, not a rebutting one. The defendant's character is evidence, but it is not strong enough evidence to override the fingerprint evidence. This indicates that the epistemic terrain is more intricate than discussions about epistemic authority recognize. The intricacy matters because it affords reason for pessimism that there will be a one-size-fits-all answer to the question of how to exploit an expert's epistemic superiority. It also suggests that rather than black boxing an epistemic superior's reasons and blindly following her lead, we should look further.

The preemption view disincentivizes personal epistemic improvement. Why should a student learn physics or history or auto mechanics if he is never going to achieve sufficient mastery to be an epistemic authority in that field? His own epistemic accomplishments will always be preempted, since epistemic superiors abound. On an exam where he is asked to explain (that is, give reason to think) that p, ought he answer, "That's what my epistemically superior, prize-winning professor believes"? Preemption diminishes the value of finding out or figuring out things for oneself. One should simply identify an epistemic superior in each area of interest and accept whatever the authority holds. Such a strategy is epistemically and morally irresponsible.

Preemption also seems to make it presumptuous to ask one's epistemic superiors for reasons. Why should they adduce evidence for p if their epistemic superiority is itself a conclusive reason for inferiors to believe that p? If there are epistemic authorities, I am probably the world's authority on Nelson Goodman's philosophy. Still, I cannot invoke my standing to close down an argument or settle a question about Goodman. If, for example, someone objects that Goodman's talk of features violates his commitment to nominalism, I cannot silence the debate by simply declaring that she is wrong. Nor can Penrose silence debate about gravitational collapse by declaring that because he is the world's authority on the topic, he gets to settle the issue. Regardless of our levels of expertise, it is incumbent on us to address the substance of the challenges. But on Zagzebski's view, if I in fact am the world's authority on Goodman, my opinion is not just a reason but a conclusive reason for others to adopt my views about Goodman. And if Penrose is the world's authority on gravitational collapse, his opinion is a conclusive reason for others to adopt his views about gravitational collapse. That is not to say that our views are true. Rather, it is to say that, if we are authorities and have no epistemic peers, they are immune to challenge.

These considerations suggest that at least one of our epistemic goals is to deepen our understanding. Arguably, acceding to authorities extends our range, giving us a larger number (or proportion) of true beliefs. But since blindly acceding to experts seems uncongenial to education and to reason giving, it impedes our understanding of the systematic supports for our beliefs. If my only reason for believing that p is that I have it from an authority, I do not know what to make of it or what to do with it. It is just another bit of trivia. Information that I glean solely from an epistemic authority is shallow. By replacing my reasons for believing that p or for believing that $\sim p$, I lose valuable information both about the phenomena and about my own grasp of the phenomena.

Still, I am dependent on others. Even if the information I receive is shallow, I cannot do without it. Here, the difference between authority and expertise becomes important. Following Goldman (2018), I said earlier that experts provide epistemic resources to novices. The issue is what sort of resources these are. I suggest that they consist in evidence about evidence as well as evidence about the adequacy of evidence, methods, standards, and criteria. Even when I am ignorant of what another epistemic agent's reasons are, the information that he believes that p is (except in cases where I consider the believer incompetent) epistemically significant. My recognition that Ralph, whom I do not consider incompetent, believes that p is evidence that p. It may not be much evidence, but it is some. My recognition that Tim, whom I consider an epistemic peer, believes that p is stronger evidence. My recognition that Sydney, whom I regard as my epistemic superior, believes that p is even stronger evidence. Depending on my assessment of the level of Sydney's relevant epistemic accomplishments, his believing that p may be quite strong evidence for p. So if I have no antecedent views as to whether p or already am inclined to think that p, my case for p is considerably strengthened when I learn that Sydney believes it. If I antecedently believe that $\sim p$ (or am inclined to believe that $\sim p$), the news that peers like Ralph disagree should give me pause. And the news that an expert like Sydney disagrees should give me strong reservations. That others believe that p is higher-order evidence that can strengthen or weaken first-order evidence (see Feldman 2014). Minimally it is evidence that others agree with the conclusions I draw on the basis of my first-order evidence. Their agreement, although it need not make my conclusion more probable, does make it more robust. In many cases, it suggests that others have additional evidence that I have not taken into account. If I have reason to think that there is more or different evidence that bears on my belief, I should take this into account in assessing the acceptability of my current belief.

The news that an epistemic superior disagrees with me should not be decisive. It provides a reason, often a strong reason. Sometimes it is a strong enough reason that I should defer to the expert—suspecting that his evidence, whatever it is, is strong enough to settle the issue. But recall the advice about getting a second opinion. Sometimes, the news that a superior disagrees with me is evidence that I should look further. Rather than replace reasons I have

176 Chapter 7

access to with reasons that I have no access to, I should perhaps consider how my belief that ~p is woven into the fabric of understanding. What does it support? What supports it? How strong and resilient are these supports? How confident should I be about them? I should also consider what becomes doubtful if I abandon my belief that ~p. What reasons, criteria, methods, and standards lose their mooring?

The tension between my views and those of the expert can pay additional epistemic dividends. I concede that the butler stole the spoons, given the expert interpretation of the fingerprint evidence. But that raises further questions. Perhaps it leads me to reconsider my assessment of the butler's character and/or my overall competence in character assessment. Perhaps it prompts a reevaluation of the evidence I tend to rely on in making such assessments. Should I still consider it good evidence? I might hold fast to my assessment of his character and ask how a person of good character might find himself in such dire straits that he is driven to steal the spoons. I might think more generally about how character and circumstances interact to account for behavior. And so forth. These avenues of inquiry are open to me, so long as I do not set my original reasons aside. The advancement of understanding often involves investigating why we went wrong when we went wrong. It is impeded if we simply disregard our previous reasons rather than asking what insights they still afford.

Acceptance or rejection of a proposition brings with it a cascade of epistemic consequences. So does suspension of judgment. The news that an epistemic superior disagrees with me calls into question my belief and my basis for belief. But to call something into question is not to supply the answer. So, what should I do in the face of expert disagreement? It depends.

Conclusion

I have argued that testimony is a matter not merely of conveying information but also of taking responsibility for the epistemic adequacy of the information conveyed. The testifier represents herself as suitably knowledgeable to shoulder the burden. In accepting testimony, the recipient concedes the testifier's expertise. He recognizes that she is in a better epistemic position with respect to the content than he is. But he does not grant her preemptive authority. He does not sacrifice his epistemic autonomy, for he retains the right to ask for reasons and to rescind his acceptance if he finds them wanting.

8   Reasonable Disagreement

A Sign of Error?

Normally, we take it that we know rather a lot—the date of the Battle of Gettysburg, the atomic number of gold, the score of last week's football game, the name of the King of England. We have reasons—often quite good reasons—for believing the things we take ourselves to know. Evidence mounts, eventually reaching or surpassing the threshold required for knowledge. But usually our reasons are less than conclusive. They just render our conclusions highly likely. Some of the reasons may themselves be epistemically insecure. Evidence may be misleading or unreliable. A source we trusted may have been misinformed. Perhaps a consideration that holds generally does not hold in this case. Given our epistemic vulnerability, it is no surprise that others disagree with us about many of the things we take ourselves to know. What should we make of that? Typically philosophers assume that where parties disagree about a matter of fact, at least one of them has made a mistake. I will argue that that is false. Because judgment involves choices from within ranges of epistemically acceptable alternatives, neither all disagreements nor all disagreements between epistemic peers rest on mistakes. This is not to say that the conclusions of both parties are true. It is to say that neither party made an error in coming to her conclusion. If this is right, it reveals something important about the precariousness of our epistemic condition and something about the value of responsible disagreement.

The standard question with respect to disagreement is how epistemic agents ought to respond. Should we hold fast to our opinions, adjust our confidence levels, or revise our commitments in the face of disagreement? When the question is framed so schematically, the alternatives seem stark:

A. Be steadfast: Hold on to your convictions, disregarding disagreement.

B. Conciliate: Moderate your confidence in the face of disagreement. Either suspend judgment about the issue under dispute or lower your credence in your conclusion.

C. Concede: Abandon your position. Admit that your opponent is right.

We need not think that the same stance is appropriate in every case. Some disagreements may call for steadfastness, others for conciliation. In yet others, the best thing to do is concede. Let's look at some cases.

Hal is aware that he is pretty clueless about ornithology and rightly considers Val quite knowledgeable. He thinks that all goldfinches look alike. Val disagrees. She thinks goldfinches display sexual dimorphism. In the face of their disagreement, Hal probably ought to concede that Val is right, or at least considerably lower his credence or suspend judgment. He recognizes that she has expertise that he lacks. Their situations are not symmetrical. The disagreement affords Val no grounds for revising her opinion. Since Hal can hardly tell a sparrow from a hawk, his opinion need not give her pause. It is reasonable for her to remain steadfast. Ceteris paribus, when an expert and a novice disagree, the novice should back off. He should either conciliate or concede. The expert, however, can reasonably hold fast.

What if parties to a disagreement are epistemic peers? Neither is more knowledgeable than the other. Here their level of competence matters. Bill and Jill are on the verge of failing introductory chemistry. Jill thinks that methane is an organic chemical; Bill thinks that it is inorganic. Given the tenuousness of their grasp of the topic, they should probably suspend judgment and look it up. If an agent's grasp of an issue is sufficiently weak, it should be unsettled by the news that anyone, even an equally ignorant peer, disagrees.

Cases like these are easily handled. Disagreements among competent epistemic peers are trickier. What should a responsible, competent epistemic agent do on learning that a peer with the same evidence, background assumptions, reasoning abilities, and epistemic incentives disagrees?[1] In what follows, I will assume that the peers under discussion are competent with regard to the issue in dispute. Conceding is reasonable only when an epistemic agent believes that her interlocutor has considerably greater relevant expertise or considerably better evidence than she does. Where peers disagree, there is no reason to be concessive. In peer disagreements, an epistemic agent should either stick to her guns or conciliate. Kelly (2005) advocates steadfastness. If

Nell considers herself to have been responsible in forming her opinion, she should dismiss her peer Mel's opinion, concluding that he must be in error. Christensen (2007) advocates conciliation. Nell should lower her credence or suspend judgment, concluding that one of the two is in error, but it is not clear who. Either way, it is widely agreed that at least one has made a mistake—miscalculated, overlooked evidence, ignored base rates, reasoned fallaciously, or committed some other cognitive gaffe. Whatever the precise failing, where peers disagree, at least one of them did something cognitively culpable. As Sidgwick says, "If I find any of my judgments, intuitive or inferential, in direct conflict with some other mind, there must be error somewhere" (1981:342).

Some disagreements, such as Christensen's (2007) restaurant case, clearly exhibit this profile. When competent agents who are equally good at mental math disagree about how to split the bill, at least one of them miscalculated. The obvious remedy is to suspend judgment temporarily, do the calculation publicly, and figure out the right answer. Similarly, if they disagree about the date of the Battle of Gettysburg. That too can be easily checked. Pretty plainly, parties to the dispute should suspend judgment, pull out their smartphones, and google the answer. Both questions are easily resolved. To suspend judgment and check the answer takes little time and effort. Conciliation in such cases is relatively cost free.

But many disagreements lack easy resolution. There is no feasible way to backtrack and find an error. Sometimes the information that would settle the dispute is simply not to be had. Ann and Dan, lifelong friends with good memories, disagree about who won the third-grade spelling bee when they were in school together. No records were kept of this momentous event. Nor are they in touch with any of their former classmates. That being so, their disagreement seems fated to remain unresolved. In cases like this, the protagonists agree about everything relevant except the fact at the heart of the dispute. Someone misremembers, but there is no way to determine who.

In other cases, however, there is no error. That is not to say that both claims are true but rather that both opinions are products of epistemically acceptable, even epistemically impeccable, reasoning from epistemically acceptable, or even impeccable premises. Although one of the protagonists arrived at a false conclusion, neither made a mistake. I want to focus on such a case.

Suppose an animal horn with holes pierced in it is found among the artifacts in a Neanderthal settlement.[2] It appears enough like a primitive flute that, had it been found in a prehistoric *Homo sapiens* settlement, paleoanthropologists would have no reservations about calling it a flute. But according to current theory, Neanderthal brains were not complex enough to create music, nor were their hands dexterous enough to finger a flute. Jen and Ken, both eminent paleoanthropologists, disagree about whether the artifact is a Neanderthal flute. Jen thinks it is; Ken thinks it is not. Their disagreement may be irresolvable. Given the paucity of evidence about Neanderthals, there is no expectation that anyone will ever come up with sufficient evidence to settle the matter conclusively. Moreover, something significant is at stake here. The verdict will contribute to the ongoing debate about how and in what respects late Neanderthals differed from early *Homo sapiens*. If Neanderthals were capable of making music, a host of related hypotheses are seriously off the mark.

Jen and Ken are equally knowledgeable and competent, and they reach or surpass the threshold for being qualified to judge the matter under dispute. They have the same evidence and the same ability to assess the evidence. Both consider the issue significant. Neither is cavalier. What should they do? Conceding has no appeal. Neither has reason to think that the other is in a better position to assess the matter. But both conciliation and steadfastness look like live options. What is to be said for them?

The case for steadfastness: Jen knows that she has considered the matter carefully and judiciously and has taken account of all the available evidence. She knows that she would have no reservations about her conclusion if she relied only on the first-order evidence, her background assumptions, and the inferences she draws from them. She rightly considers Ken her intellectual peer in paleoanthropology. So, she is surprised that Ken does not agree with her. If she is steadfast, she does not, however, consider his disagreement a reason to rethink her position or moderate her confidence level. She concludes that, in this case, Ken must have made a mistake. Maybe he was careless in his reasoning; maybe he overlooked some relevant factor. If he had thought the matter through as carefully as she did, she thinks, he would have come to the same conclusion. Ken's disagreement gives her no reason to change her mind. Kelly (2005) argues that the fact that a peer disagrees is evidence. Huemer (2011), Foley (2001), and Henderson et al. (2017) maintain that there is a first-person privilege that serves as a tie breaker. Although they diverge over the grounds for that privilege, they think that it suffices to justify sticking to one's guns.

Reasonable Disagreement 181

Ken's situation is symmetrical. He too recognizes that he thought the matter through carefully and judiciously. Being equally steadfast, he can only conclude that Jen, although normally quite solid on such issues, has made a regrettable error. Her disagreement gives him no reason to change his view either. If they remain steadfast, Jen and Ken may be permanently at loggerheads. Since each is inclined to dismiss the other's opinion as erroneous and to have no incentive to examine the other's reasons, neither is in a position to learn from the other. The disagreement gives them no reason to rethink the issue. Evidently, the steadfast are *intellectually arrogant*. Faced with a disagreement, a steadfast epistemic agent immediately concludes that the other party must be wrong. They are *dogmatic*. They do not think that a peer's disagreeing gives them any reason to rethink their position or change their mind. They are, moreover, *disrespectful* of those they consider intellectual equals. Indeed, we might wonder whether they really *do* consider their opponents intellectual equals if they are so quick to claim superiority the moment a disagreement arises.

The case for conciliation: We've seen that Jen has thought through the matter carefully and judiciously. She would have no reservations about her conclusion if she relied exclusively on the first-order evidence, her background assumptions, and the inferences she draws from them. But Ken, whom she considers her epistemic peer, disagrees with her. She has no reason to think that Ken was any less careful and judicious in his approach to the issue or that he was cavalier in his reasoning or interpretation of the evidence. One of their beliefs must be false. But given that they are peers, there is currently no way to tell which one. She therefore suspends judgment, hoping that further evidence will emerge that will decide the issue between them. Conciliators are *intellectually humble*. Their epistemic stance is *fallibilist*. Jen thinks she is right but acknowledges that she could be wrong. Conciliators are *respectful* of their epistemic peers. As far as Jen can tell, Ken's view has as much going for it as hers. There is no basis for choosing between them.

It might seem that this shows that when epistemic peers disagree, they should conciliate. Certainly, conciliators seem more congenial than the steadfast. But conciliation can be costly. If Jen and Ken take their disagreement as a reason to suspend judgment about whether the artifact is a Neanderthal flute, neither of them is in a position to know. Similarly, if they lower their credences to below the threshold for knowledge. They have backed off from belief about the matter. If Jen's first-order evidence was actually sufficient for her to know, then by suspending judgment, she sacrifices knowledge, for knowledge requires belief. One of them is surely right. Either the artifact is a Neanderthal flute or it is not. If both hold fast to their opinions, then one of them has a true belief. And if the one with the true belief has sufficient evidence that bears non-accidentally on its truth, one of them knows. If Jen is right and sticks to her guns, she knows or at least justifiably believes that the object is a Neanderthal flute. So, it seems, the steadfast are capable of preserving knowledge, while the conciliatory willingly give it up.

Steadfastness, however, instantiates the so-called dogmatism paradox that Kripke identifies (2011:43):

If S knows that p, then p is true.

If p is true, then any evidence against p is misleading.

If S wants to retain her knowledge, she should disregard misleading information.

So S should dogmatically disregard any evidence that seems to tell against p.

This argument is perfectly general. It applies to first-order evidence, as well as to evidence that S might glean from the discovery that T disagrees with her. It is not strictly a paradox, but it yields an epistemically disheartening tension. It divorces knowledge from rationality. Rationality requires that epistemic agents be open-minded—that they not intentionally blind themselves to evidence, even if that evidence might turn out to be misleading. So if Ken's disagreement is evidence that Jen is wrong, Jen should not disregard it. Although the steadfast can retain knowledge or highly justified belief, they do so by shirking their epistemic responsibility. It is just by luck that the evidence that might be gleaned from taking the disagreement seriously turned out to be misleading. Given that Jen and Ken are epistemic peers, it is just by luck that Jen (rather than Ken), by being steadfast, retains knowledge.

Conciliators, however, may appear spineless. The moment they learn that a peer disagrees, they back off from full acceptance. It seems epistemically ill-advised, even a bit cowardly, to move too quickly from "I might be wrong" to "I should immediately abandon my position." If this is what conciliation requires, it is not an attractive option. There is, however, some slack in the conciliationist position. Conciliators hold that on learning that a peer disagrees, an epistemic agent should either *suspend judgment* or *lower her credence*. Suspending judgment is a considerable sacrifice, at least if that means

Reasonable Disagreement

having no opinion, taking no stand, or thinking one is unjustified in one's view of a matter. Given the range of peer disagreement, most of us would have to suspend judgment about a lot. But it may be reasonable to conclude that we should lower our credence in our conclusions. Conciliators could still be fairly confident that their conclusion holds, but they would have to admit that they are not fully confident and are not entitled to be fully confident about it. This might result in a loss of knowledge. Whether it does depends on where the threshold for knowledge is and how far above the threshold an agent's commitment originally was. Arguably, it often *should* result in a loss of knowledge. The agent would not think that she is, or should consider herself to be, utterly clueless about the topic, nor would she or should she consider herself spineless in backing off. But she would and should have her confidence in her conviction shaken. She might nevertheless remain pretty sure of the conclusion.

Conciliators gain more than they lose. We saw that the steadfast are dogmatic. They dismiss their opponent's opinion as mistaken without further investigation. If steadfast Ken is wrong, he is likely to remain wrong, since unless he encounters an epistemic superior who agrees with Jen, he takes himself to have no reason to rethink the issue. The existence of the disagreement gives him no incentive to investigate the matter further. Conciliators have such an incentive. Jen and Ken have the same evidence and the same background information. On learning that Ken disagrees with her, Jen has reason to ask, "What does he see that I do not?" She has reason to go back and examine the evidence, the background assumptions, her inferences, and, if she has access to them, his. Has she overlooked something? Has she reasoned fallaciously? Is there an illuminating perspective on the issue that she missed? Conciliators are in a better position than the steadfast to find the mistake that led to the disagreement, assuming, of course, that there was a mistake.

Faultless Disagreement

But the assumption that disagreement always rests on a mistake is itself a mistake. The issue is why the peers disagree. Even in the idealized scenario, where peers are identical in multiple relevant respects, there can be a variety of reasons. Although by stipulation the peers have the same evidence, background assumptions, inferential abilities, and epistemic motivation, there is no reason to assume that they use these resources in the same way. If they do not, they may arrive at different verdicts. And there may be nothing clearly mistaken about the way either of them proceeds. In such cases, what is at issue is not just whether p but also how one ought to determine whether p. I will urge that our epistemic situation is more complex than is usually recognized. Peer disagreements bring this complexity to the fore.

Disagreements occur in a multidimensional space of alternatives. Along a variety of axes, epistemic requirements fix ranges within which acceptable verdicts must lie. But the requirements are not sufficiently fine-grained to assure uniqueness or to provide a decision procedure for differentiating among judgments that fall within the range of acceptability. If parties disagree because they make different choices within a range of epistemically acceptable options, neither makes a mistake. Where an agent comes down depends to a considerable extent on which alternatives she chooses. Had she made other choices within the acceptable range, she would have reached a different verdict.[3] Epistemic peers are not epistemic clones.

By stipulation, epistemic peers have the same evidence. But it does not follow that they assign the same weight to different bits of evidence. The disputed artifact was found in a cave in Slovenia, in an area known to have been populated by Neanderthals. Jen considers this highly significant, thinking it constitutes fairly strong evidence that the artifact was crafted by a Neanderthal. Ken attaches less weight to this datum, noting that the perforated horn could have been left in the cave by a wandering *Homo sapiens* or been brought there by a Neanderthal, having been crafted elsewhere by a *Homo sapiens*. Ken does not deny that the location where the object was found is evidence, he just considers it relatively weak evidence.

Peers are said to have the same inferential abilities. But they need not have the same reasoning styles. Ken tends to reason analogically and to credit analogical arguments. Jen, although equally adept at analogical reasoning, prefers abductive inference, considering analogical arguments rather loose. Ken saw his dog pick up a ball in the yard this morning and carry it over to her bed. He buttresses his belief that the artifact might have been transported to the Neanderthal site by relying on the analogy with his dog's behavior. Jen thinks that the best—simplest, most plausible—explanation of an artifact's being found in a Neanderthal site is that it was crafted by a denizen of that site—namely, a Neanderthal. Ken is more skeptical of abductive inference, thinking that such linear reasoning tends to blind one to available alternatives.

Reasonable Disagreement 185

In characterizing a newly found object as a Neanderthal artifact, paleo-anthropologists compare it with objects that are already characterized as such. If it is similar enough to acknowledged Neanderthal artifacts in relevant respects and different enough from prehistoric artifacts deemed clearly to be not crafted by Neanderthals, it is likely to be incorporated into the precedent class and henceforth classified as a Neanderthal artifact. Otherwise it is apt to be rejected. But the community recognizes that the current precedent class contains some items that were wrongly classified. That is, it contains misleaders. Ken and Jen differ over which items are misleaders. This leads them to different verdicts about this artifact. Jen credits scored bits of flint that are thought to have been used as tools. She thinks that if the Neanderthals had the cognitive and manual dexterity to knap such tools, they had the cognitive and manual dexterity to pierce holes in an animal horn to make a flute. Ken doubts that the gashes in the flint were intentionally produced. He suspects that what Jen considers intentional scores are in fact natural striations, amplified over eons by friction, wind, and water. So the evidence afforded by the flint tools leads Jen to think the perforated horn is a Neanderthal artifact, while Ken harbors doubts. It may be obvious that a precedent class contains misleaders. Perhaps various elements are in tension with one another in the sense that if one is a Neanderthal artifact, another probably is not. Even so, it need not be obvious which members of the class are the misleaders. Peers then can disagree because they differ over which members of the (admittedly somewhat messy) precedent class they consider worthy of trust.

A subject's background information about a topic consists of all the information she has that directly or indirectly bears on that topic. Although epistemic peers, by stipulation, have the same background information, they need not draw on the same bits of information or assign the bits they draw on the same weight. If they do not, background information plays different roles in their reasoning. Contemporary anatomical theory suggests, albeit weakly, that organisms with thick phalanges are relatively lacking in fine-motor control. Because Neanderthals had thick phalanges, Ken thinks that they would have been unable to finger a primitive flute. Jen is not convinced. She considers the anatomical theory sketchy and its bearing on the case slight.

Many epistemologists follow William James in holding that our overarching epistemic objective is to believe as many truths as possible and to disbelieve as many falsehoods as possible (1948:99–100). They typically then focus exclusively on believing truths. But James's formula involves a proportion (see Riggs 2003). If an agent is epistemically risk averse, she will set a high threshold for acceptance. Perhaps she will accept only hypotheses that are 95 percent probable on the evidence. If she is more daring, she sets a lower threshold, perhaps 90 percent. Both thresholds comply with the requirement that an acceptable conclusion must be highly probable. The risk-averse agent will accept fewer truths and fewer falsehoods than the more risk-tolerant agent. But so long as the "highly probable" requirement is met, neither is epistemically irresponsible. Jen is willing to accept a larger measure of epistemic risk than her mentor, Ben, is. She concludes that the artifact is a Neanderthal flute because, having taken all of the previously discussed factors into account, she deems it 92 percent probable that it is. Ben, being more conservative, suspends judgment because he demands a probability of 95 percent. He agrees with all of Jen's assessments. He simply does not think that they yield a high enough probability.

Evidently peers can differ along a variety of epistemically relevant axes: the weight to assign to evidence, the standards of acceptability, the identity of misleaders, the relevance and importance of various bits of background information, their favored styles of reasoning. To limit the cast of characters, I described Jen and Ken as differing along all these axes. Ben was introduced to show that even with widespread agreement in numerous other respects, epistemic agents can differ over thresholds of acceptability. The factors I've mentioned vary independently and can point in different directions. Someone who sets a high threshold on acceptability could easily favor or disfavor analogical reasoning. Along each axis, all three parties hold reasonable views. None is making an obvious mistake. Nor is it plausible that any of them is making a subtle mistake. We should not expect there to be a precise numerical value for how risk averse one should be nor should we think that there is an algorithm for identifying misleaders, assigning weights to different bits of evidence, or assessing the weight to attach to an analogical argument.

Perhaps this just shows that the epistemologists who set the criteria for peerhood should include more constraints. They should insist that epistemic peers also agree about the weight of evidence they assign, the cutoff on acceptability, the reasoning strategies they actually use, and so forth. That just postpones the predicament, for the same arguments could be given on each side with a bit more precision. Once we control for all the differences I indicated (and probably others that I overlooked), the familiar alternatives reappear. We could, by fiat, insist that epistemic peers be epistemic clones. They disagree about nothing except the verdict. Then we would revert to the standard way of thinking about peer disagreements. But the stipulation just sweeps the complexity of the epistemic situation under the rug. Even if Joe and Moe do not qualify as epistemic peers because they assign different permissible weights to the evidence, there is a difficulty. The fact that acceptability is a matter of falling within a given range rather than landing on a particular point needs to be accommodated. Whether or not they qualify as peers or merely near peers, we get back to the original question: Should they be steadfast or conciliatory?

The rationale for the original constraints on peerhood was to focus attention on disagreements that are epistemologically telling. As we have seen, not all are. Drawing on legal terminology, let us say that a factor is dispositive when it settles how a disagreement should be resolved. In some cases, imbalances in levels of expertise are presumptively dispositive. Once we recognize that one party is a novice and the other is an expert or that one party has a relevant cognitive asset and the other a corresponding deficit, it is usually clear how the disagreement ought to be resolved. But in disagreements between epistemically competent peers, there are no (even presumptively) dispositive disparities. To be sure, if we think the current criteria have omitted some dispositive differences, we can augment the criteria. Elsewhere I suggested, for example, that besides being equal in the respects standardly recognized, epistemic peers should be equally well educated in the relevant disciplines (Elgin 2010). But the grounds for disagreement that we have canvassed here seem different. In each of the respects we looked at, it is not at all obvious that either of the protagonists is in an epistemically superior position. We can determine how many additional false positives and false negatives will result from taking 90 percent rather than 95 percent as the threshold for acceptability, but that by itself does not tell us where the line should be drawn. Setting a precise threshold for how probable a conclusion should be or how much weight should be attached to a given bit of evidence seems arbitrary. Moreover, it is apt to be counterproductive. Philip Kitcher (1990) argues that the scientific community may best serve its collective epistemic ends by, at any given time, supporting a range of conflicting views. When there is a nonnegligible chance that a currently disfavored view is true, if the members of the community want to believe what is true, it would be premature to foreclose inquiry. Multiple ways of balancing the value of believing truths against the disvalue of believing falsehoods are often permissible. There is no reason to think there is a single optimal balance.

I suggest that we look at the availability of these sorts of disagreements as epistemic assets. The diversity of ways in which peers (as originally characterized) can still reasonably disagree should sensitize us to the complexity of a situation and the epistemic opportunities that are available through taking different perspectives on it.

Judgments may be sensitive to choices among epistemically acceptable alternatives. Had an agent made different choices within one or more acceptable ranges, she would have justifiably come to a different conclusion. This highlights the precariousness of our epistemic condition. That precariousness is independent of disagreement. The dependence of Jen's opinion on choices would obtain even if she knew nothing about Ken's opinion—indeed, even if Ken had no opinion on the matter. This might suggest that my discussion just underscores our fallibility. Despite the agent's best efforts, she still could be wrong.

Perhaps so. But disagreements among competent peers can be epistemically fruitful. If properly investigated, they provide resources for focusing fallibilism. If parties dissect their disagreement, they can discover how the various epistemically responsible choices affected their verdicts. Jen and Ben, for example, might come to appreciate that the basis for their disagreement has nothing to do with Neanderthals per se. Rather, it is about how great a risk of error paleoanthropology ought to tolerate. Given mutual respect, disagreement might also provide peers with an incentive to rethink their choices. Ken might be prompted to reconsider whether he would view the matter differently if he revised his opinion about which elements of the evidence class are misleaders. Jen might rethink her doubts about the strength of analogical arguments. Even if all parties end up endorsing their previous positions, their stance vis-à-vis them is different. They appreciate that it is a disagreement not just (or perhaps even mainly) about the maker of a particular artifact but also about epistemic methods and standards that affect how one ought to determine the status of the artifact.

Open-mindedness is a propensity to entertain alternative points of view. Not all points of view merit attention. None of our protagonists has any reason to think that the artifact in question is a space alien's cleverly disguised slide trombone. Nor should they. A competent peer's disagreement effectively certifies that a perspective is worth taking seriously. It raises the question: What is to be said for his way of looking at things? Given that he is a peer, the presumption is that there is something to be said for it.

Disagreements in Philosophy: A Case Study

The picture I have sketched about the value of disagreement has an additional payoff. It vindicates the way we do philosophy. As standardly described, neither the steadfast nor the conciliatory really engage. The steadfast agent ignores the disagreement and holds fast to whatever she already believed. The conciliatory agent immediately backs off. (She may then recalculate or recalibrate, but it is open to her simply to suspend judgment or lower her credence without doing anything else.) Few if any philosophers exhibit either sort of behavior. If anyone were so bold as to simply dismiss a worthy opponent's disagreement out of hand, concluding that he must be wrong somewhere but there is no premium in attempting to figure out where, he would not be doing philosophy. Philosophy is not dogmatic. Someone who immediately suspended judgment on learning that a peer disagreed would not be doing philosophy either. Philosophy is not spineless. Faced with disagreement, philosophers engage. We attempt to uncover the basis of the disagreement and assess its merits. We look for shared or unshared presuppositions, common or diverging conceptions, flaws in our own or our opponent's arguments. Some of us may be sufficiently convinced of our views that we are quietly sure that our interlocutor has made a mistake somewhere, but we still take it that we need to figure out where. Others may be more open-minded about whose mistake it is and even whether there is a mistake. In any case, we assume that it is our responsibility to take the disagreement seriously.

The peer disagreement literature presents disagreement as epistemically undesirable. Apparently, all things considered, we'd prefer that folks agree. As we have seen, learning that a peer disagrees with you is supposed to prompt you to suspend judgment, lower your credence, or dismiss your peer's conviction as somehow flawed, even if you can neither identify the flaw nor explain why you think she is the party in error. Philosophers do none of the above. A distinctive feature of analytic philosophy is that, although we marshal the strongest arguments we can devise, we do not really expect others to agree.

190   Chapter 8

Nor should we be particularly daunted or even disconcerted when they don't (see Goldberg 2013a). That, I suggest, raises questions about what we think we are doing. To underscore the puzzle, let's look at some curious cases.

An Anecdote

Shortly after Nelson Goodman and I published *Reconceptions* (1988), a workshop on the book was held at the University of Bielefeld. The workshop had the typical format. Someone presented a paper. One of us gave a short reply. Then the floor was open for a general discussion. One issue we discuss in *Reconceptions* concerns the identity of literary works. The inscriptionalist criterion we advance—that sameness of spelling is the criterion of identity for a literary work—has the unattractive consequence that translations are not strictly instances of the work they translate. We devised something that, in computer terminology, might be called a patch. It handles the problem but in a graceless, seemingly ad hoc way. At the workshop, Wolfgang Heydrich (1993) presented a paper that respects our nominalist scruples and supplies an elegant alternative.

Here the important point is what happened next. We replied, "He's right. His alternative is lovely. It is much better than ours. We wish we had thought of it." Because everyone apparently agreed with our assessment, no one knew what to do next. Once we said he was right, there seemed little left to discuss. Our collective befuddlement may have been exacerbated by the technical nature of the issue. In effect, Heydrich said: here's a glitch; here's a fix. Probably in other cases, an argument that got (and merited) universal agreement would open avenues for fruitful discussion. "If, as we all agree, Professor Z has conclusively demonstrated that p, that opens the door to q, r, and s." Still, the reactions at the Bielefeld workshop underscored the fact that the standard expectation in philosophy is that participants to an exchange will disagree: we will raise pointed objections, ask challenging questions, identify and spell out the significance of weaknesses. *Philosophers are typically ill-equipped to give or take "yes" for an answer.*

We might contrast philosophy's practice with another genre in which one person tells others what to think or do: the sermon. The topic of a sermon and a philosophy lecture could be the same; the lecture and the sermon might even be delivered by the same person. Suppose Reverend X is also a philosopher. This week he gives both a sermon and a philosophy lecture on, say, the nature of virtue. Perhaps he even gives the same talk both in church

Reasonable Disagreement

and in the lecture hall. (I have been told that Bishop Butler did this.) Still, there is a difference. In church, there is no Q&A. There is no formal, real-time opportunity for members of the congregation to dispute what is said. They are supposed to sit quietly, take it in, and, if need be, revise their beliefs and reform their behavior accordingly. After the philosophy lecture, however, auditors are expected to raise objections—and the speaker is expected to reply. Regardless of the esteem with which they hold the lecturer, philosophers do not and should not let his words go unchallenged. Disagreement is expected.

Lit Reviews

Many philosophy papers begin with a review of the literature. The practice goes back to Aristotle, who often began his discussions with "Let us examine the opinions of the many and the wise." Like Aristotle, we review the important positions on the subject under discussion. (We may ignore the opinions of the many, but we consider it obligatory to address the opinions of the [allegedly] wise.) Occasionally, we accept and build on what another philosopher has said. But even there, we are apt to intimate that the position we build on regrettably did not go far enough. A literature review in philosophy typically consists of rehearsing extant positions and highlighting their inadequacies. That done, we go on to present our own (which, naturally enough, does not suffer from the flaws we highlighted). The literature review sets the stage. It frames our discussion by explaining how and why we disagree with others who have worked on the same topic. This is different from the natural sciences where lit reviews typically emphasize why the works being cited are a good basis to build on.

Hiring Decisions

David Lewis (2000) notes that philosophy departments typically ignore the question of whether the candidate's philosophical views are true. If the goal of the department and the university is the advancement of knowledge, this practice is epistemically costly. Ceteris paribus, someone with false beliefs on the topic he plans to devote a considerable portion of his professional life to is less likely to advance knowledge than someone with true beliefs on the topic. The reason, Lewis says, is this: "To the extent that a researcher is guided by false doctrines, he is liable to arrive at new and different false doctrines, since he will choose them partly to cohere with the doctrines he held before" (2000:189). Although this is rather obvious, hiring committees in philosophy routinely disregard it. Our hiring practice is odd. We purport to seek the truth yet bracket the question of whether the candidate's views are true. If we sought a medical opinion or advice about how to fix the furnace, we would presumably restrict our informants to those we had reason to think held true beliefs about the topic. Minimally, we would eschew those whose beliefs we thought were manifestly false. Why are philosophy hiring decisions different?

Lewis suggests that the practice is the product of a tacit treaty. Since there is no assurance that the majority in any given philosophy department believes the truth, there is a chance that if the truth value members ascribe to a candidate's views were considered relevant, the resulting hiring decisions would entrench falsehood. Hence we tacitly agree that all parties will willingly forgo the best outcome (bringing it about that the department is dominated by those with true philosophical beliefs) to block the worst (bringing it about that the department is dominated by those with false ones) (2000:198). Maybe so. But when the result of the treaty is that the department hires someone whose views a member regards as false, the person being hired will be her colleague for the foreseeable future. So they will disagree. Lewis's argument suggests that this is a regrettable consequence that we all have to live with. Below, I will suggest a more positive spin.

Response to Fallibility

As I mentioned earlier, Philip Kitcher (1990) argues when there is a significant chance that a currently popular scientific theory is false, the community has sound epistemic reasons to ensure that alternative positions remain viable. This requires that adherents of those positions get some proportion of the jobs in the field, a measure of financial support to enable them to continue their investigations, and opportunities to present their results and be assured of a fair hearing. If the members of the community want to believe only what is true, Kitcher argues, it should not foreclose inquiry prematurely. Presumably such open-mindedness is restricted to scientifically plausible alternatives. Kitcher is not recommending that the National Science Foundation fund the research of contemporary followers of Thales who are bound and determined to demonstrate that, the history of science notwithstanding, everything is water.

Kitcher's argument applies to philosophy at least as much as to science. Maybe more. However confident we may be about the epistemic status of currently accepted scientific theories, we should probably suspect that our favorite philosophical position, regardless of its popularity, stands a good chance of being false. As Richard Fumerton points out, "most philosophical views are minority opinions" (2010:109; see also Bourget & Chalmers 2014). On any issue of philosophical significance, there is nothing close to consensus. Not only should we therefore refrain from foreclosing inquiry into currently unpopular but plausible rival views, we should provide venues where the merits of different positions, even implausible ones, can be compared and contrasted—that is, venues where disagreements can be aired and taken seriously. To be sure, we don't want to spend our careers fending off kooks and nuts. So we will have to figure out how to exclude them. But the criterion for exclusion should be something other than the kookiness or nuttiness of the theses being proposed.

Required Courses

Many philosophy departments require their students to take courses in the history of philosophy. They strong-arm their students into spending considerable time and intellectual energy studying positions that their mentors reject as clearly false—positions that they both hope and expect the students will also reject. We require our students to study Plato, even if we are convinced that the forms do not exist. We require them to study Spinoza, even if we are confident that there's more than one thing in the world. We require them to study Kant, even if we consider it obvious that transcendental idealism is a nonstarter. I do not suggest that this is bad practice. In fact, I think that studying the great philosophers of the past is vital. I will say more about why shortly. Still, there is something odd about expecting ourselves and our students to spend a lot of time studying positions that we reject—and reject for good reason. We actively disagree not only with the views of our contemporaries but also with those of our forebears. And we can't disagree with them if we don't know what they are.

Philosophy's curious institutional arrangements and practices seem to indicate that we consider responsible philosophical disagreements valuable. What can that value be? Before addressing that, we should consider what it is to hold a philosophical position.

Commitment

Many philosophers assume that to hold a philosophical position is to believe it. This strikes me as incorrect. To believe that p entails believing that p is true.

Although some philosophers believe the theories that they favor to be true, most, I suspect, do not. If one believes that a theory is true, one ought to believe that it will never justifiably be rejected. Any objections raised against it are misleaders. *Never?* Is it remotely plausible that two hundred years from now, even our best philosophical theories will be accepted exactly as they stand? Even if you are, for example, so convinced of epistemological externalism that you believe that in two hundred years some version of externalism will dominate epistemology, is it plausible that it will be the very version you or Alvin Goldman or Hilary Kornblith worked out? Here, I suggest, the pessimistic meta-induction is sound. If we look at the history of philosophy, we do not find a body of received truths that were never subject to revision. It seems unjustifiably arrogant to think that *my* theory and *my* arguments for it (or Goldman's or Kornblith's) are so powerful that they will escape the fates that befell Aristotle, Descartes, Kant, Frege, and whatever intellectual heroes you want to add to the list. Nor is the pessimistic meta-induction the only ground for doubt. Fumerton maintains that he probably shouldn't believe his own theory given how many of his very smart friends and colleagues do not (2010:109). He is not alone in this. To believe that p entails believing that p is true, not just believing that p is somewhere in the general neighborhood of the truth.

To deny that philosophers believe their theories is not to deny that philosophers are and should be committed to their theories. It is to say that their commitment is not a matter of belief. We are, and think we should be, committed to our philosophical views. What is the nature of that commitment?

Goldberg suggests that our attitude should be:

> *Attitudinal speculation*: One who attitudinally speculates that p regards p as more likely than $\sim p$, though also regards the total evidence as stopping short of warranting belief in p. (2013b:284)

Theories are complex. They consist of a multiplicity of interwoven commitments. There are any number of dimensions along which a philosophical theory could be wrong. Holding that a theory is more likely to be true than false is excessively optimistic. To think, for example, that Korsgaard's constructivist ethics or Sosa's virtue epistemology or Chakravartty's scientific realism is more likely to be true than to be false is to think that it is unlikely that there is a mistake anywhere in the theory.

Reasonable Disagreement

Another problem can be seen if we assess a view against its rivals. Suppose a consequentialist recognizes that the plausible rivals to his theory are deontology and virtue theory. (To keep things simple, we will ignore his attitudes toward rival consequentialist theories.) He might well hold that consequentialism is more likely to be true than either deontology or virtue theory. But he might still assign the probabilities as follows:

consequentialism—40 percent;

deontology—30 percent;

virtue theory—30 percent.

In that case, he does not think the theory he is committed to is more likely to be true than to be false. He does think it is more likely to be true than any one of its rivals (Barnett 2019:114). Barnett suggests that Goldberg would do better to accept

> *Speculation**: One who attitudinally speculates* that p regards p as the likeliest option (given some set of options), though also regards the total evidence as stopping short of warranting belief in p. (2019:115)

Barnett argues that even with this modification, Goldberg faces a problem. Peer disagreement provides higher-order evidence that the position an agent espouses is unwarranted. So the vast number of competent philosophers who disagree with consequentialism supplies enough evidence against it to bring its probability to less than 50 percent. And the vast number of philosophers who think that some other view, say deontology, is correct diminishes the probability that consequentialism is the likeliest alternative. The bad news is that this is not just a problem for moral theories. It holds throughout philosophy.

Barnett suggests that our attitude toward our favored philosophical position is what he calls "disagreement insulated belief" (2019:121). The position you should hold, he argues, is the position that is the likeliest option given your evidence, once you have set aside the evidence of disagreement. It is, in effect, what you would believe if you were insulated from information about what your peers believe.

There are a couple of difficulties with this suggestion. The first is that it ignores the pessimistic meta-induction. That well-supported theories in philosophy have been rejected or seriously revised in the past should give me pause if I'm tempted to think that my theory will have a happier fate.

A second worry is that insulating ourselves from disagreement deprives us of information. Even if I would justifiably believe a theory if I was unaware that others disagree, the news that they disagree seems germane. Once I learn of their disagreement, it seems irresponsible to ignore it.

To appreciate the importance of disagreement for philosophy, I suggest, a reorientation is in order. Standardly, epistemological discussions about disagreement focus on differences of opinion as to the truth value of a particular proposition. This carries over to the discussions of disagreement in philosophy. Such discussions are cast as differences of opinion as to whether to believe, for example, externalism or internalism, consequentialism or deontology, scientific realism or constructive empiricism, as though the issue is what particular proposition to endorse. But such generic positions admit of multiple, divergent theories. A confirmed reliabilist presumably holds that a belief qualifies as knowledge only if it is reliably connected to the facts. Still, opinions diverge over which connections are reliable and how reliable they have to be. There is plenty to disagree about within the reliabilist camp. Theories are intricately interwoven networks of commitments. Disagreements in philosophy concern what network of commitments is best on balance. They are not typically reducible to disagreements as to the truth value of a particular proposition. I'll have more to say about this shortly.

I suggest that philosophy is a quest for understanding rather than for knowledge. If we want to understand the moral realm or the epistemic realm or the metaphysical realm, we want to grasp how a variety of considerations interweave to constitute a tenable take on the topic. The question then is not whether externalism is true but whether, and if so, how and how adequately, externalism affords insight into the epistemic realm.

Rather than belief (or belief minus something), we should focus on *epistemic acceptance*. To accept a commitment, theory, or network of commitments is to be willing and able to use it as a basis for inference or for action when one's ends are cognitive. The commitments that comprise an epistemic network are not all judgments of fact. They include norms, standards, methods, rules, even prospects. Nor is inclusion in a system in reflective equilibrium expected to ensure truth. A system's being in reflective equilibrium only ensures that the system is as reasonable as any available alternative in the epistemic circumstances. As a result, someone can justifiably accept a theory while recognizing that it is likely eventually to be superseded. She can justifiably accept a theory while recognizing that it has rivals that are equally tenable. She can even justifiably accept a theory while recognizing that it is a bit of a long shot. Then its claim to equal tenability would be based on an assessment of its promise as compared with the promise of its rivals. An evidentialist, for example, might think that the current support for her position is somewhat worse than the support for reliabilism but consider its prospects better. She might think this, and be justified in thinking it, on the basis of an assessment of the trajectories of the improvements in the two positions over time. In that case, she would hold that even if reliabilism has a slight edge today, evidentialist theories are improving at a faster rate. (Improvement would be measured, presumably, in terms of increasing ability to solve epistemological problems.) Helen Beebee (2018) argues, on the basis of the ubiquity of irresolvable philosophical disagreement, that we should take the goal of philosophy to be equilibrizing—that is, finding a view that is as good, all things considered, as any of its rivals. This does not require or yield truth or knowledge. But it does make sense of philosophical practice and the reasons we value it. Equilibrizing is not a concession to the regrettable fact that we do not have knock-down arguments. The reorientation that Beebee and I advocate puts us in a position to see how philosophical disagreements can be valuable.

Bet Hedging

Kitcher (1990) and Lewis (2000) make a strong case that epistemic diversity is valuable. By keeping alternatives to a received view alive, we hedge our bets. The issues under investigation are difficult; available resources evidently afford considerably less than conclusive reasons. So it makes sense to incorporate into our disciplinary practices a recognition that, despite the evidence in favor of a particular position, it still might be wrong.

A division of cognitive labor is an efficient way to sustain diversity. Rather than expecting each of us to master the pros and cons of each of the alternatives and fair-mindedly argue "on the one hand, . . . ; on the other hand, . . . ," it may be preferable for one party to develop and, to the best of her ability, argue for one position, while another develops and argues for an alternative. This strategy provides an incentive to tolerate and even encourage the development of positions at odds with one's own. It does not, by itself, provide any reason to value disagreement per se. To reap the benefits of cognitive diversity, it might be best simply to allow for parallel tracks. One group champions dualism; another, monism. One champions nominalism; another, Platonism. One champions internalism; another, externalism. Each can proceed in isolation from the other. The downside is that such a strategy would impede the opportunity to learn from one another.

Owning up to our fallibility is not as helpful as we might hope. To acknowledge that the position put forth in a thirty-page article or a 250-page book might be wrong somewhere does not give any indication of where or how or why it might be wrong. Disagreement can play a diagnostic role. It provides focus by pitting positions against one another. This is why the disputational style of philosophy's Q&As (or the style they should display) pays epistemic dividends. A would-be questioner who simply muses about what he considers related issues is unhelpful. So is one who takes the opportunity to spout off about his own pet theory rather than addressing the position that has just been presented. So is one who announces that the position just has to be wrong but can't identify any defect. Such self-styled questioners distract from the goal of the practice: to put the author of the paper and the auditors in a better position to understand and evaluate the claims being made. But a questioner who carefully articulates a worry—something of the form, "if you say that *a*, doesn't that commit you to *b*?"—pinpoints a locus of concern. She does not simply announce that something *might* be wrong somewhere or even that it *must* be wrong somewhere. We already knew the first and may strongly suspect the second. She indicates where she thinks a problem lies. She might, of course, make her case by appealing to her own theory. But the appeal should be relevant and apt. If it is, then by looking at the argument from her perspective, we may discern something worrisome.

Actual disagreement is not required to perform this function though. Valuable discussions often involve challengers who play devil's advocate—asking, "What would you say if *p*?" when they do not for a moment countenance *p*. Inasmuch as philosophical positions are supposed to be defensible against bizarre as well as plausible alternatives, this is a familiar and worthwhile practice. Still, actual disagreement is apt to dig deeper. Someone who responsibly disagrees with a position is likely to have thought harder about the issue than a mere occasional devil's advocate. She has probably subjected the position that generates her objection to scrutiny. She is intellectually invested in it. So, ceteris paribus, her objection will probably have more traction than one a devil's advocate can formulate on the spot.

Reasonable Disagreement

Much of the literature on disagreement poses the problem as though the believer knows nothing more than the brute fact that a peer disagrees. But normally we know who disagrees, what they disagree about, and why. Even just knowing the identity of the philosopher who disagrees with us is informative. If, for example, I learn that Miranda Fricker disagrees with me, I worry about whether I've been insensitive to epistemic injustice; if I learn that Sophie Horowitz disagrees with me, I worry about whether I have violated Bayesian constraints. Once I am privy to the content of the disagreement, I can consider why the objector holds an opposing view. What does she see that I do not? What is she focusing on that I am not? The same goes for knowing what the locus of disagreement is or why it is taken to be important. If I insulate myself from disagreement, as Barnett advocates, I may never know. One epistemic contribution of responsible disagreement, then, is to pinpoint errors, vulnerabilities, or potential problems in a position. That is not the only benefit.

Advancing Understanding

A critical question concerns the basis for the disagreement. Philosophical disagreements are typically not like Christensen's (2007) restaurant case. If they are at all serious, no one has made an obvious mistake or omission. Sometimes, although I suspect rather rarely, there is a disagreement about a simple matter of fact or logic. More often though, the grounds are different. We have seen that peers can disagree along a variety of epistemically relevant axes: the weight they assign to evidence, the standards of acceptability, the identity of misleaders, the relevance and importance of various bits of background information, the favored styles of reasoning, the acceptability of different trade-offs. If, for example, a consequentialist maintains that the good is prior to the right, while a deontologist maintains that the right is prior to the good, those priorities are likely to figure in the assessment of arguments, in judgments about particular cases, in whether or to what extent an unforeseen outcome tells against a moral judgment, and so forth. The consequentialist and the deontologist adopt different orientations toward the moral realm. Each perspective highlights some seemingly relevant factors and occludes others (see van Fraassen 2008:37–39). Each provokes questions that the other does not. A deontologist probably need not worry much about unforeseeable consequences; a consequentialist has to. A scientific realist has to concern herself with the condition of Schrödinger's cat; a constructive empiricist might not.

Disagreements then are not merely additional reasons to worry that you might be wrong. They expand your epistemic range, disclosing previously unrecognized powers and limitations of your own view. They underscore that the truth is never the whole truth. They reveal that there are potentially important factors that have been sidelined, omitted, unappreciated, or ignored. They sensitize you to weaknesses or limitations of your position. Even if your argument is solid, it may be vulnerable if the background assumptions are not precisely as you take them to be. Disagreements acquaint you with relevant alternatives and the considerations that might favor them. They put you in a position to recognize the range of considerations your position depends on—for example, how strong your modal commitments are, what types of evidence you can draw on and what weight that we should attach to it, perhaps what boundaries you set on your theory. (Is it an ideal theory, or is it supposed to reflect real world constraints? Should you be concerned about the 'fact that it holds only under a limited range of actual circumstances?) These features are epistemically valuable because they enable you to better understand your own theory, the alternatives to it, and the topic it bears on.

A theory at odds with yours may have attractive features that your theory lacks. A nominalist thinks there is no chance that Lewis's realism about possible worlds is correct. But if she temporarily sidelines her reservations and seriously entertains it, she may appreciate that it has the resources to solve a variety of problems that her metaphysically austere position cannot. She learns something about the problem space she is working within, the questions she would like to answer, and the resources she has (or lacks) for answering them. She can recognize that Lewis's theory highlights the importance of distinctions that her theory elides. By taking it seriously, she can discern perhaps regrettable shortfalls in her own position, the costs that have to be paid to eschew metaphysical extravagance. Even though she remains convinced that nominalism is more tenable overall, she also recognizes that it doesn't do everything she might want. Maybe by studying Lewis's position, she can identify features that she can export. Perhaps, for example, armed with a suitable theory of fiction, a fictionalist about possibility could adopt some of Lewis's insights or strategies. To figure out if this is so, she needs to locate her disagreement with Lewis. What exactly does she disagree with? What features of the theory are peripheral to their disagreement?

A theory, even a correct theory, is not an intensional replica of reality. In theorizing, we decide what factors are important, how the domain is to be partitioned into kinds, at what grain and in what vocabulary the data are to be characterized, what factors can be set aside or treated as of only marginal significance, and so on. There are trade-offs. One theory favors precision; another, breadth. One regards epistemic accessibility as paramount; another cheerfully allows that that the truth may be unknowable.[4]

When we structure a domain we draw lines, stipulating that some things will be counted alike, others as unlike. Metaphysically, our stipulations are bound to be correct. Every collection of objects consists of items that are somehow alike. If nothing else, they are alike in being members of that particular collection. Most likenesses, although real, are utterly idle. There is no reason to attend to them. The critical issue in structuring a domain is to draw lines that group together items whose likenesses matter and to differentiate items whose differences matter. This holds not just for objects but also for principles, norms, relations, and standards. It is unsurprising that there should be controversy about where lines should be drawn. Often critical disputes in philosophy stem from disputes about what issues are central.

Given the number of dimensions along which there are choice points, we should expect a variety of positions to emerge. They will highlight different aspects of the phenomena. Consequentialists, for example, highlight the fact that in acting we aim to produce a particular outcome. They maintain that the act should be assessed in terms of the outcome. Deontologists recognize that we are hostage to fortune. We may act with the best will and the best plan in the world, and still our actions come a cropper. That being so, they maintain, since ethics is concerned with blameworthiness, it should focus on the moral character and the intentions of the agent, not on the outcome of her actions. Such divergences can provide a foothold for potentially fruitful disagreements. Rather than simply hurling invectives at one another, we might learn from one another. We can ask, "What is to be gleaned by seeing things from their point of view?" Even learning that a seemingly plausible theory is untenable can be informative (see Mill 1978:49–50). The discovery that the appeal to natural kinds won't solve the new riddle of induction, absent an argument—rather than a mere intuition—that *green* rather than *grue* is the more natural kind, reveals something about the depth and complexity of the problem.

Responsible Disagreement

For disagreement to be epistemically valuable, it must be responsible. I said earlier that we must block nuts and kooks if we are going to get anywhere. How are we to do that if not by announcing that a disagreement grounded in realism about possible worlds or in the prospect that all emeralds are grue is too kooky to be taken seriously? I suggest that rather than assessing the content of the theses, we need to consider the epistemic contours of the disagreement.

Is the epistemic agent who disagrees competent? This is not a matter of what degrees he has but of whether he displays and draws on an understanding of the topic and the ways it might fruitfully be approached. Are his arguments cogent? To decide this involves an assessment of the tenability, appropriateness, and use of background assumptions, evidence, and modes of reasoning. Is he conscientious? That is, does he take his epistemic responsibilities seriously in this matter? Has he done his homework? If he neglects information that is readily accessible or overlooks relevant and available evidential or logical resources, there is reason to doubt that his disagreement will reveal anything epistemically significant. The requirements of competence and conscientiousness provide sufficient reason to, for example, dismiss the objections of anti-vaxxers, since they either do not understand or do not respect the evidence that has discredited the contention that vaccines cause autism or implant nanobots. Their position has already been justifiably rejected. So unless and until they can provide new, more plausible evidence, their disagreement can, for epistemic purposes, be ignored. On the other hand, we should probably take seriously the view of an epidemiologist who maintains that because the flu virus has mutated, this year's flu vaccine will be relatively ineffective. Similarly, someone who dismisses skeptical arguments out of hand on the grounds that there are no malevolent demons need not be taken seriously. Someone who argues that inference to the best explanation blocks demon arguments probably deserves a hearing.

Nevertheless, we should take pains not to dismiss disagreements cavalierly. Openness to new ideas is crucial. A position that looks kooky may turn out to be insightful. Repeated reconsideration of a previously repudiated

Reasonable Disagreement

position is fruitless. So is devoting time and effort to assessing a contention that is not backed by reasons. But since a community of inquiry must be open to innovations, the mere fact that a suggestion is nonstandard is not grounds for dismissal. Nor, as we have seen, is the fact that the agent who ventures it is not a member of the current epistemic elite. The issue turns on the bona fides of the idea, not those of the agent who proposes it. Some kooky ideas turn out to be correct.

Responsible disagreement needs to be relevant and focused. Cato the Censor is said to have ended every speech he gave in the Roman Senate, thundering, "Carthage must be destroyed!" For all I know, he was right about Carthage. But there were, no doubt, many debates in the Senate where his point was irrelevant. If he voiced his disagreement with current foreign policy in a debate over, for example, a plan to construct a temple or fix an aqueduct on the grounds that it did not address the danger that Carthage posed, then his disagreement probably should have been disregarded. Similarly, a philosopher obsessed with the hard problem of consciousness who disagrees with every speaker who fails to solve the problem—even if that speaker is discussing truth-maker semantics or the role of beauty in aesthetics—should probably often be disregarded. For him to insist, "But you haven't solved the hard problem of consciousness!" although true, would not be particularly useful. Unless he can show that to say something important about the role of beauty in aesthetics requires taking a stand on the hard problem of consciousness, the news that a particular aesthetician did not solve or even address the problem is not significant.

Responsible disagreement should be respectful. It should conform to the principle of charity. The reason is not just that doing so would be polite. By giving the strongest available reading of a position before disagreeing with it, you disclose something that cuts to the core of the issue. A disagreement that bears only on a shallow or uncharitable construal does not shed much light on the central issues under discussion. The problem it points to either is peripheral or is easily solved.

Fruitful disagreements illuminate the phenomena, our current understanding of the phenomena, and our current resources for addressing the relevant issues. They reveal something worthwhile that we were unaware of or insufficiently attentive to. They may sensitize us to weaknesses or vulnerabilities in our approach. They prompt rethinking.

Curious Practices Redux

This discussion sheds light on the epistemic value of some of philosophy's curious practices. Peer agreement typically does little to advance understanding. That like-minded folks agree with me may assuage my insecurities, giving me confidence that I haven't made a glaring error. Maybe it affords evidence that I have a relatively stable base to work from. But it does not do much to push my thinking further. It does not point me in any particular direction. A fruitful disagreement, on the other hand, advances understanding. It can highlight aspects of my theory and its relation to the phenomena that I had not properly attended to. Suppose, for instance, I construed theoretical knowledge as a conjunction of propositions, each of which has a high probability, and construed the growth of knowledge as a matter of adding new conjuncts. Finnur Dellsén (2021) disagrees, pointing out that it is a principle of probability theory that the probability of a conjunction of mutually independent claims decreases with each additional conjunct. As a result, each additional independent conjunct whose probability is less than 1.0 lowers the probability of the whole. My account, as it stands, faces a serious problem. Maybe I'll decide to scrap it; more likely, I will modify it— perhaps by ceasing to construe theoretical knowledge as a conjunction or by rejecting the idea that probability is the measure of tenability.

The advantages of fruitful disagreement can also arise from agreement. I said above that agreement among like-minded people does not advance understanding much. Surprising agreement is different. If, for example, a scientific realist and a constructive empiricist find themselves agreeing about the best way to interpret Newton's bucket or find themselves with the same reasons for being bewildered about Einstein–Podolsky–Rosen, they may come to understand that (and with luck why) the issues raised by these thought experiments transcend their respective stances toward unobservables. Such agreement, because it was unanticipated, discloses something new and potentially epistemically valuable about where their positions do and where they do not diverge.

Why do we require our students to study the history of philosophy? Why do we turn to our predecessors as often as we do? Not surprisingly, Aristotle provides the answer. Recall his appeal to the many and the wise. In the lead-up to his definition of eudaemonia in the *Nicomachean Ethics*, he lists a number of views and says, "Now some of these views have been held by

Reasonable Disagreement

many men and men of old, others by a few eminent persons; and it is not probable that either of these should be entirely mistaken, but rather that they should be right in at least one respect or even in most respects" (§1098b25–30). We look to our forebears not merely for inspiration or to wholly endorse or wholly reject what came before but to cull from their accounts ideas that strike us as worth expropriating, at least in part. Although we think they are wrong, we suspect that they are not entirely wrong.

Something similar holds for literature reviews. Because philosophers often frame their positions by juxtaposing them with presumptively plausible alternatives, we use peer disagreement to triangulate. By sketching the current state of play, perhaps with acknowledgement of the historical trajectory that got us here, we map the terrain of the territory within which we will work. We position ourselves by reference to those we disagree with. Indeed, we sometimes define the problem we are dealing with by disagreeing about the ways the problem has previously been defined. This would be fruitless if we thought they were dead wrong. In *De Anima*, Aristotle says, "It is necessary . . . to call into council the views of those of our predecessors who have declared any opinion on this subject, in order that we may profit by whatever is sound in their suggestions and avoid their errors" (§403b20–23). We look at the works of our (partly mistaken) predecessors and peers with an eye to deciding what might be worth importing, perhaps with modifications, into our own theory. Rather than just dismissing them as wrong, we seriously investigate what is wrong with them. By figuring out where and why we agree and where and why we disagree, we benefit from the efforts of others. We don't then have to reinvent the wheel; nor, if we are lucky, do we fall into the same traps they fell into.

Hiring practices foster understanding in ways that Lewis does not acknowledge. This, I suspect, is because he focused on truth and knowledge and did not consider how the quest for understanding might be different. Suppose a department sought to advance understanding in metaethics.[5] It would make very good sense to hire Peter Railton—a confirmed moral realist—and Allan Gibbard—an equally confirmed expressivist. The reason is not mainly because the department wanted to hedge its bets or diversify its intellectual portfolio. Rather, the expectation is that, over many years, Railton, Gibbard, and the students they draw to the department will fruitfully disagree. They will push one another to respond to objections they never would have thought of on their own. The department also has an incentive to hire Liz

Anderson, who will raise objections if, for example, the metaethical theorizing departs too far from everyday moral life. Even if one of them arrives at the truth about metaethics, it is doubtful that he would have achieved such a refined, well-supported, comprehensive, and well-defended understanding without the years of fruitful disagreements with colleagues and students. We learn from one another. We benefit from being pushed to rethink, strengthen, and refine our views. In philosophy, the push often comes in the form of disagreement.

I have argued that peer disagreement is an asset in philosophy; it is not something to shy away from. I've suggested that this is reflected in our practices. Still the question arises, when confronted with a case of peer disagreement, should we suspend judgment, lower our credences, or hold fast to our positions?

I've suggested that most philosophers do not actually believe their theories. So strictly speaking, lowering our credences is not an available option. Still, it might seem that we could lower our level of confidence in our network of commitments, which would amount to much the same thing. This sleight of hand is no help. Most of us probably expect that our views will be supplanted. This seems to indicate that whatever we are confident about, it is not that our views, exactly as they stand, are correct.

Should we then suspend judgment? Similar problems arise here. In being committed to a philosophical position, what are we supposed to be judging? What is it we are supposed to suspend? Acceptance is a matter of being willing to use as a basis for inference or for action when our ends are cognitive. To suspend acceptance then is to deprive ourselves of a resource. Sometimes doing so is entirely reasonable. Once you become convinced, say, that naive set theory leads to paradox, you should withdraw your commitment to it and refuse to use the axiom of unrestricted comprehension as a premise in cognitively serious inferences. This is an extreme case. It is a matter for outright rejection, not suspension. Often, however, our epistemic resources seem worth preserving. The mere fact that, for example, a smart deontologist disagrees is not enough, and arguably should not be enough, to prompt a consequentialist to cease reasoning from consequentialist premises. To suspend, it seems, is to refrain from using either a commitment or its denial in reasoning about a topic. That seems to be a heavy price to pay. Suspension of consequentialist commitments and their denials, or deontological commitments and their denials, would eviscerate the resources for moral reasoning.

The third alternative is to remain steadfast. As it is usually characterized, remaining steadfast requires not only retaining your commitments but also believing that your peer has made an error in opposing them. I suggest that we can hold to our commitments without making any negative assessment of our peer. On my view, to hold to one's commitments is to continue to consider them worthy of being used as a basis for inference and for action when one's ends are cognitive. This does not entail, or even suggest, that one's peer is wrong to retain her commitments.

There is considerable epistemic benefit in this alternative. We learn more about our own commitments and their vulnerabilities. We learn more about alternatives to our position and the trade-offs that favor one side or another. Figuring out exactly where and why you disagree with David Lewis or Judith Jarvis Thomson is an intellectually rewarding experience that gives you resources for strengthening your own theory. Taking well-supported philosophical disagreements seriously is intriguing, informative, and fun. This is not to say that philosophers should never concede or conciliate. Since we already expected disagreement, our confidence need not be diminished merely because we discover that others disagree. What matters is the content of the disagreement. In some cases, it ought to convince us to withdraw a commitment. Russell's (1971) objection to Frege's Basic Law V was conclusive. Logicism is untenable. Concession is called for. In others, it might prompt serious rethinking. Vogel's (1990) car theft case might lead you to rethink your commitment to epistemic closure. Perhaps we should conciliate. In yet others, we are justified in holding fast. For now, I remain committed to a nonfactive account of understanding, even though a nontrivial number of very smart philosophers disagree with me.

Although I have been focusing on philosophy, I suspect that such an account of disagreement holds for systematic inquiry generally. Most systematic inquirers probably expect that their findings will eventually be supplanted. Most probably recognize that judgment calls lie behind some of the decisions they make. If this is right, then in the context of systematic inquiry, responsible disagreement, rather than being grounds for dismay, should be recognized as an epistemic asset.

If we recognize that disagreement does not always indicate that someone has made an error, we can introduce a further stance—one that I think philosophers take almost by default. Parties might remain *committed* to their own position, considering it a good basis for inference and action when their ends are cognitive. Nevertheless, they recognize its vulnerability. This recognition is not idle. It requires them to be open to, and perhaps even to seek out, emerging counterevidence and counterarguments. This blocks Kripke's dogmatism puzzle. It gives them a strong incentive to strengthen and stabilize their position and to protect themselves should it turn out that some of the considerations they countenance are wrong. It undermines complacency and fosters intellectual respect. Like the steadfast, the committed retain their convictions; unlike the steadfast, they do not dismiss disagreement or denigrate those who disagree. Rather, they exploit responsible disagreement, treating it not just as a reminder that they might be wrong but also as a probe to tease out exactly where and why they might be wrong.

If the parties to the disagreement entertain one another's position seriously and respectfully—if they keep a genuinely open mind to the possibility that there is something significant to be said for it—they may come to appreciate weaknesses in their own position as well as strengths in their adversary's. This may lead one to concede that the other party is correct. Alternatively, it may lead her to shore up her own position so that it can deflect his objections. A third possibility is that together they craft a position different from the one either of them started with, or revise their methods, standards, or criteria of acceptance. There is no assurance that they will ever agree. Nor, as Kitcher (1990) argues, should we think that we would be better off if we could devise criteria that ensured agreement in either the short or the long run. But by entertaining respectable alternative positions and appreciating their merits, we understand our epistemic situation better. Moreover, if we appreciate the basis of our disagreement, we are in a better position to introduce appropriate safeguards. Given our fallibility, it is in general wise to take into account the possibility that a well-supported opinion might still be false. We hedge our bets, take out insurance, make backups, diversify our portfolios, carry umbrellas in the recognition that our confident expectations do not always pan out. Attention to disagreement can pinpoint the places where safeguards are most called for. In cases like the ones we've discussed here, the safeguards in question might involve introducing caveats or highlighting the ways a conclusion is based on assumptions that could legitimately be challenged. It might also involve efforts to stabilize one's position by showing that the conclusion is not too dangerously dependent on a controversial choice. If Jen could arrive at her conclusion without strongly depending on the disputed

Reasonable Disagreement

shards of flint or without dismissing analogical arguments, her case would be stronger. If she sees how Ken's argument works against her, she can marshal resources for the defense. When we take disagreements among competent peers seriously, we gain a richer, more focused appreciation of our epistemic predicaments. Whether we decide to remain committed or to conciliate, we are in a better position to appreciate the nature, scope, and insights to be gleaned from and about our epistemic vulnerability.

9 Beyond the Information Given

Introduction

We have seen that testimony regularly augments our epistemic resources. Via testimony, we glean not just factual information but also warrant for that information. We know a vast array of facts that we did not learn firsthand. Electrons have negative charge; Napoleon was defeated at Waterloo; *Homo sapiens* evolved from *Homo erectus*; Paris is the capital of France. The literature on the epistemology of testimony takes for granted that the teaching of such facts consists in testimony (see Adler 2002:138–161). If so, we came to believe—indeed, to know—that electrons have negative charge because a teacher who knew that fact testified—that is, told us—and we properly uptook his testimony. This, epistemologists say, is obvious. Indeed, according to Nickel and Carter (2014), such classroom-gleaned knowledge is a paradigm case of testimonial knowledge. In this chapter, I argue that, paradigmatic or not, this is incorrect. Teaching and testimony are distinct. Although both testimony and the teaching of matters of fact involve conveying information, their epistemic profiles diverge. In making my argument, I draw on familiar pedagogical practices. I take it for granted that those practices pretty much align with their goals.

Teaching as Testimony?

As we have seen, testimony can impart complex information conveying logical structure, spatiotemporal order, and dependence relations. It may therefore seem obvious that the teaching of factual material consists in testimony. The teacher delivers a factual lecture, testifying about a topic such as the role of the cotton gin in the perpetuation of American slavery or the life cycle of the periodical cicada. In so doing, she states facts that she knows. The students, being recipients of her testimony, come to believe the facts imparted. This, it is held, is what occurs in successful teaching in disciplines like history, the social sciences, and the natural sciences. For now, let us call these fact-focused disciplines. I will have more to say about this characterization below. If this conception of teaching is correct, then so long as the lecturer knows the facts she imparts and the students have no defeaters, when they believe what she says, they know it. The testimony model is held to apply across grade levels. The fourth-grade teacher imparts simple facts; the university professor imparts subtle, complex, nuanced facts. But in all cases, the knowledgeable lecturer testifies, and successful students uptake that testimony.

On standard views, successful testimony transmits epistemic entitlement. This is consonant with epistemic reliabilism (see Goldman 1999).[1] If the recipient acquires the information from a reliable informant and has no defeaters, he knows. It might seem that the testifier must have the level of entitlement she transmits. She cannot, via testimony, bring it about that her recipient knows that p if she does not know that p. Nor can she bring it about that her recipient is justified or warranted in believing or accepting p if she is not justified or warranted in believing or accepting it. She might, of course, spark an insight in a listener, enabling him to know because he, independently, had the justification she lacked. But this would not be testimonial knowledge, for in such a case, although she prompted him to know, she did not convey knowledge to him. This seems obvious. You cannot transmit what you do not have.

In a vignette called Creationist Teacher, Jennifer Lackey challenges this view. Her protagonist, Stella, is a devout fundamentalist Christian who teaches fourth grade in a public school. Stella believes in the truth of creationism and the falsity of evolutionary theory. But being a responsible teacher, she considers it her duty to teach her students the theory that best answers to the evidence, not the theory that is supported only by her own personal faith. She therefore gives a reliable, age-appropriate lecture in which she asserts "Modern-day *Homo sapiens* evolved from *Homo erectus.*" Stella does not believe—hence does not know—it. But, it seems, her students who take her at her word come to harbor the true belief that that *Homo sapiens* evolved from *Homo erectus*. It is at least very tempting to say that they come to know it (Lackey 2008:48).

Beyond the Information Given

Lackey maintains that this vignette discredits the view that testimonial knowledge is grounded in the transmission of beliefs. She contends that whether Stella believes is irrelevant; what matters is the reliability of the content Stella conveys. If Lackey is right, it is possible for testimony to impart knowledge even if the testifier does not know.

Federica Malfatti (2019) extends Lackey's challenge, considering whether testimony that is not believed can be a vehicle for conveying understanding. In Malfatti's vignette, Stella's counterpart, Lilith, is a climate science professor who does not believe in anthropogenic climate change. Nevertheless, via Lilith's teaching, her students come to understand and accept the reality of anthropogenic climate change. They are able to appeal to it to explain, for example, why glaciers are receding and sea levels are rising. The issue, it seems, concerns conveying or imparting or bringing about epistemic entitlement, whether that entitlement is a matter of warrant, knowledge, or understanding.

Robert Audi, writing before Lackey and Malfatti, maintains that students who acquire their beliefs from teachers like Stella in fact do not know. Such a teacher is a hypocrite, he suggests. She would teach whatever the school wanted her to teach, regardless of what she thought about its epistemic standing. That being so, her students ought not take her word. Alternatively, he suggests, a teacher like Stella might be unduly credulous. In that case, she suffers from a cognitive dysfunction that renders her epistemically untrustworthy. Either way, the students fail to know, for they ought not take her word (Audi 2006:30). If Audi is right, Stella's students do not know, and Lilith's students do not understand.

Since Stella and Lilith are imaginary, we can credit them with any traits we like. But we ought not disparage their moral or intellectual character too quickly. Perhaps they are constructive empiricists (see van Fraassen 1980). They restrict, and think they ought to restrict, their teaching to theories that are, as far as they can tell, empirically adequate. Let us call such theories "evidentially adequate." An evidentially adequate theory is one that is, on the available evidence, empirically adequate. Since the theories in question fill the bill, Lilith and Stella can in good conscience teach those theories. Each opts to teach the evidentially adequate theory of her subject that is most widely accepted by the relevant scientific community or examination board.[2] It is, they think, epistemically permissible to teach any evidentially adequate theory, and it is practically beneficial to their students to teach the one that is generally accepted in the relevant venues.

One might wonder whether this stance is sufficient to bring it about that the students know that *Homo sapiens* evolved from *Homo erectus* or understand why anthropogenic climate change is causing glaciers to recede. Perhaps not. But the problem verges on a skeptical worry, for according to constructive empiricism, empirically adequate theories accommodate all the evidence there is. That evidence is not conclusive. Still, it is widely recognized that we can know and understand things on the basis of less than conclusive evidence. A constructive empiricist who favors the alternative that happens to be true may be in as good a position to know as any other epistemic agent who believes a truth on the basis of less than conclusive evidence. Stella, of course, does not know, for she does not believe. But fourth graders are apt to be naive realists. Perhaps her lesson gives them sufficient reason to believe. And since the claim is true and grounded in the evidence, if they believe on the basis of her teaching and have no defeaters, they know. Although the relation of understanding to truth is less straightforward, it is plausible that Lilith's students understand if they reflectively endorse the thesis that anthropogenic climate change is real and that it accounts for phenomena like the rise in sea levels and the melting of glaciers. Audi's reservations then need not discredit Lackey's and Malfatti's claims.

Stella and Lilith are not pedagogical outliers. When teachers simplify, as they often do, they do not, strictly speaking, believe the content they teach. Philosophy professors go further. We regularly teach theories that do not even come close to what we believe. It might seem that in doing so, we implicitly embed our claims within an "in the theory" operator in the way that a literature professor embeds a claim like "Hamlet was indecisive" within an "in the story" operator. Then, with "in the theory" tacitly assumed, a professor who does not endorse Platonism might responsibly utter, "The forms are prior to and independent of their material instantiations."

Following de Regt and Dieks (2005), Malfatti (2019) distinguishes between understanding a theory and understanding a range of phenomena via that theory. It might seem that in teaching a theory that we do not accept, we simply aim to convey an understanding of the theory but not an understanding of the phenomena via the theory. Arguably this works for historians of science teaching Ptolemaic astronomy or phlogiston theory. But philosophy professors do more than explicate; we routinely mount defenses for theories we do not accept. We adduce reasons—sometimes very powerful reasons—for the views we discuss. Moreover, we draw out implications (both positive

Beyond the Information Given 215

and negative) that go beyond the texts. Nor do we restrict ourselves to reasons found in (or tacitly assumed by) the text or its author. Sometimes we strengthen a position by introducing resources that were unavailable to the thinker we are discussing. Thus, for example, we might draw on twentieth-century discussions of the failure of logicism to strengthen Kant's argument that "$7+5=12$" is a synthetic a priori truth. In so doing, we maintain that there is more to be said in favor of the theory than its author was in a position to say. A professor who does not believe that synthetic a priori truths exist could in good conscience make such a move. She might go on to attempt to dismantle the Kantian position. Even so, a student could be impressed with the move and unconvinced by the dismantling. Then, if mathematical truths really are synthetic a priori truths, the student would come to know or at least reasonably believe that they are. Giving the strongest possible explication and defense of the theories we discuss is what philosophy professors are supposed to do. The principle of charity requires it.

We might defend this strategy by maintaining that we build up a theory we reject in order to knock it down. In that case, if my students emerge believing on the basis of my lecture that mathematics consists of synthetic a priori truths, I failed, at least with the "knock it down" step, for I believe no such thing. But it is not obvious that I failed. In any case, if this is our rationale for mounting defenses of theories we reject, it is not a very good one. Why should we bother? "Here's a ridiculous view; here's why it doesn't work." This sounds like an idle exercise or a parlor game rather than a valuable pedagogical strategy. What do we think is to be gained by encouraging our students to seriously entertain views that we repudiate? As we have seen, I believe there is a good answer to this question, but it is not one that those who construe lectures as extended testimony can provide.

In any case, some philosophy professors take pains to refrain from knocking down the views they discuss in their lectures. Rawls (2000, 2008) was notoriously fair-minded in his courses on the history of moral and political philosophy. He never said, "Here's why my theory is better," even though he, if any of us, was entitled to make such a claim. If our goal is to transmit the truth, wouldn't it be preferable to simply present and argue for the view that we take to be true? Many philosophy professors hope and expect that students will reject a theory that they personally reject but defend in their lectures. Still, a student may believe it—and justifiably believe it nonetheless. Not all of us consider ourselves to have failed if she does.

Lackey's vignette is widely held to create a serious problem for belief transmission theories of testimony. Stella's students evidently emerge from the lesson knowing that modern day *Homo sapiens* evolved from *Homo erectus*, even though Stella does not. I agree that they know. But this does not call for a revision in or rejection of belief transmission views of testimonial knowledge. Nor do the epistemic achievements of Lilith's students call for a revision or rejection of belief transmission views of testimonial understanding, for the students' epistemic successes are not products of testimony. The assumption that information-conveying teaching is testimony and the assumption that the resulting learning consists in acquiring testimonial knowledge are unfounded. They do not square with our pedagogical practices or their goals.

Teaching

Much teaching involves uttering declarative sentences, and much learning involves appropriate uptake of those utterances. Israel Scheffler (1960) distinguishes between two sorts of sentences on the basis of their uptake conditions. His focus is teaching and learning content expressed by sentences. *Fact-stating* sentences, such as "Caesar crossed the Rubicon," are successfully taught if and only if the students come to believe the stated fact and retain that belief at least long enough to repeat it on the exam. Uptake seems largely a matter of receptivity. Scheffler is careful to take no stand on what facts are. All that matters for his discussion is that the information can be conveyed in a declarative sentence. Teaching via fact-stating sentences aligns well with standard views of imparting knowledge via testimony.

Norm-stating sentences, such as "Honesty is the best policy," have norms as their content. They can function in the same way as ordinary fact-stating sentences, in which case uptake consists in students' believing them and retaining that belief for a suitable period of time. But, Scheffler maintains, norm-stating sentences also admit of an active interpretation. Under that interpretation, a student has not learned the material unless she comes to display a pattern of action consonant with the norm that the sentence expresses. Thus on an active interpretation, a student who continues to behave dishonestly would not qualify as having learned that honesty is the best policy, regardless of what she wrote on her exam. She would be missing something significant.

Scheffler contends that susceptibility to active interpretation is what differentiates fact-stating sentences from norm-stating ones and what differentiates

Beyond the Information Given

the conditions on successfully teaching norms from those on successfully teaching facts. I am not convinced. I think that all successful learning involves the adoption of what Scheffler calls "patterns of action." Consider the seemingly fact-stating sentence "Rattlesnake bites are often fatal," uttered by a naturalist leading students on a field trip in the desert in Arizona. If the students just inscribed the information in their notebooks, planning to memorize it for the exam, something would have gone badly wrong. The point of the utterance was to bring it about that the students avoid rattlesnakes and appreciate why they should avoid rattlesnakes. Achieving the first goal would involve getting them to adopt a pattern of overt behavior: avoiding rattlesnakes. It would also involve a pattern of attention: looking out for rattlesnakes. Achieving the second goal would involve the adoption of patterns of inference and modes of thought. "Rattlesnake bites are often fatal" then seems to be a fact-stating sentence that admits of an active interpretation.

Not all fact-stating sentences implicate stock patterns of overt action. "Caesar crossed the Rubicon" seems not to point toward doing or avoiding anything in particular. But it does involve enabling learners to adopt certain patterns of thought: propensities to draw some inferences and eschew others, to form and entertain some hypotheses and block others, to find certain historical factors salient and set others aside, and so forth. In this respect, it is like both "Honesty is the best policy" and "Rattlesnake bites are often fatal." This indicates that it is not the content expressed by the sentence but the patterns of thought and action that the sentence, uttered in a particular context to a particular audience, activates that is important. On the agential view I advocate, all sentences bear active interpretations. To accept p is to be willing and able to use p in inference and action. It is to be equipped to do something, whether mental or physical. Teaching that p aims to equip the learner with patterns of thought and perhaps action relevant to accepting that p. If this is so, then the relation of teaching to testimony needs further investigation.

John Greco takes a more standard view. He holds that lecturing in fact-focused disciplines consists in testimony. Such a lecture is a sequence of Scheffler's fact-stating sentences. Greco recognizes that even in fact-focused disciplines, education involves more than imparting information. He attempts to accommodate the residue by assigning distinct pedagogical functions to distinct components of a fact-focused course. In particular, he maintains, science labs have a very different pedagogical function from lectures.

In science courses, he says, "The purpose of a lab component is not to learn the results of experiments, but to learn how to run experiments. That is, the purpose is to teach the kind of knowledge-how necessary to be a practitioner in the discipline" (2020:142). It is reasonable to teach the sort of know-how needed to be a practitioner in a discipline to those who intend or aspire or are in training to become practitioners in the discipline. On Greco's view then lab courses for science majors and graduate students in the sciences make sense. But high school science courses and gen-ed science courses required in many universities have lab requirements as well. Most of the students taking such courses have no intention of becoming practitioners of experimental science. This raises the question of why labs are considered cognitively valuable for such students.

On my view, the main function of the labs is not to equip students with the know-how of practitioners. It is not like teaching apprentice chefs how to make a roux or teaching fledgling bassoonists how to shape a reed. Nor is it to get students to learn the results of experiments. The function of the lab component in a science course is to exemplify the way the science produces results that are epistemically acceptable—that is, to show how and why such results afford evidence for the theories they support or evidence against the theories they discredit and to show why the norms the experiments are subject to are epistemically appropriate. Students learn what evidence can be generated for the theories they are studying, how that evidence is generated, and how and to what extent it supports the theories. They learn why intersubjective agreement and replicability of results are epistemically important. They may learn how difficult it is to generate good evidence. The understanding they gain through the course (lectures, labs, discussion sections, problem sets, etc.) is not simply a matter of testimonial knowledge, nor is the information gleaned from the course properly credited only or even mainly to the information imparted in the lectures. Labs enable students to see for themselves and show for themselves why certain theoretical claims are (or are not) acceptable. Good courses are designed to be seamless wholes. They meld lectures, discussions, reading assignments, problem sets, student reflections, term paper assignments, perhaps various types of project-based learning. They merge group work with individual assignments. This is so from the earliest grades through graduate school. The goal is that the components form a mutually supportive structure that taken together afford a broader and deeper understanding of the subject matter.

Not all fact-focused courses have labs. History courses do not. But a good history course is not restricted to imparting factual information about, as it might be, the defeat of the Spanish Armada, the demise of the Ming dynasty, or the political consequences of the Kansas–Nebraska Act. It also teaches students what sort of evidence there is for historical claims, what qualifies certain items to count as evidence, and why and to what extent such evidence is trustworthy. It teaches how to responsibly weave well-supported judgments of historical fact into an overall understanding of a historical episode. Greco might maintain that a history lecture involves some claims that convey information via testimony and other claims that perform other functions. Maybe so. A lecturer might assert, "The Kansas–Nebraska Act became law in 1854." This seems the sort of fact that students might come to know via testimony. If the students came to believe it entirely because the lecturer said so, they would have testimonial knowledge of that fact. But parsing claims on the basis of their putatively separate functions is unlikely to do justice to their educational role. An effective lecture is typically an interwoven tapestry of first- and second-order, factual, evidential, and epistemically normative claims.

Greco's conception of teaching as testimony is what Paolo Friere calls the "banking model." In the banking model, Friere maintains, teachers in effect open students' heads and deposit the information they want their students to have. They withdraw it as needed—namely, in exams (Friere 2000:72). Teachers *impose* a view of things on their students. They dictate what the students should believe. Friere's main concern is political. He argues that the banking model prevents the politically oppressed from recognizing the depth and character of their oppression. He is surely right. But the objectionability of the banking model is more extensive than he acknowledges. The model oppresses not only the politically disempowered it also oppresses the children of the powerful, for the banking model stifles thought. Rather than teaching students how to think, it stipulates what they should think. The reason they should believe that p is simply that they were told it. The objectionability is not grounded in the idea that the information imparted to the students is false or unjustified or oppressive, although it may be. The objectionability lies in its circumventing students' epistemic autonomy.

It might seem that in identifying Greco's orientation with the one Friere disparages, I overlook something crucial. Greco holds that the teachers and professors he is talking about impart truths. There is, one might think, nothing wrong with seeking to bring it about that one's students believe what is true. Friere's grievance is that the oppressors impart falsehoods to the oppressed. Indeed, they impart morally pernicious falsehoods. Obviously that is an important difference. But it is not the whole story and may not even be the main part of the story. Teachers of matters of fact want their students to believe what is true. And there are often specific truths that they want their students to learn. But they do not want the students to believe these facts simply because they were told to. Knowing the truth is, educationally, not enough. As Robertson says, "Educators should seek not merely to *transmit* knowledge, but to also put students into a position where they can, to some extent, decide what to believe. It would be a poor education that transmitted a fixed body of facts without also developing the resources for arriving at new beliefs and evaluating old ones" (2009:17; see also Scheffler 1960).

The speech act that conveys testimony is called "telling." Let us call the speech act we use in lecturing "professing." It consists in putting forth a proposition or other representation for serious cognitive purposes (see Elgin 2017:21–22). So, for example, a biology teacher professes that the Krebs cycle figures in the oxidation of glucose when she says so in her lecture. A history teacher professes that the Greeks won the battle of Marathon when he says so in his lecture. Professing is different from telling, even when the same fact is conveyed, for successful uptake is different.

As we have seen, a stereotypical case of acquiring testimonial knowledge consists in getting directions. Jed wants to know how to get to the zoo; Judy is happy to tell him. Jed uptakes that information and follows her directions. To make the case comparable to a standard lecture, let us suppose that the route is fairly complicated. After a few minutes, and a few turns, Megan asks, "Why do we turn left here?" Jed's epistemically appropriate answer is, "Judy said we should turn left on Grove Street." Jed knows where to turn because and only because a reliable informant told him where to turn. That's plenty good enough.

Compare this to the case of a student taking an exam, writing an essay, or answering a question in class. Suppose the question is:

What is the function of the Golgi apparatus?

or

What caused the extinction of the dinosaurs?

Beyond the Information Given

or

What tribe dominated the southern plains in the nineteenth century?

followed by the instruction "justify your answer." A student who gave the correct factual answer but justified it by saying "Ms. Winthrop said so" would not do well. If the lecture's function were to convey knowledge via testimony, "She said so" would be an entirely adequate justification. It would probably be an epistemically stronger justification than whatever the student could adduce by way of evidence. Evidence is merely indicative, whereas if Ms. Winthrop knows, her testimony affords a compelling—indeed (barring defeaters), a conclusive—reason to believe. To pass an exam, write an acceptable essay, or answer the question appropriately, the student is expected to display an understanding of the relevant facts and, perhaps, of the methodological and normative considerations that justify the answer. It is not enough that she can cite the expert who imparted them. She should provide evidence that bears on what makes the claim true (or true enough), not just reasons why we should think it is true (or true enough). The students are expected to provide cogent justifications to support their claims.

This suggests that the learning that is a product of teaching is subject to internalist—that is, evidentialist—constraints (see Robertson 2009). If Goldman is right that testimonial knowledge is externalist, this is an epistemologically important difference. Still, it is worth noting that merely opting for internalism about testimony would not bridge the gap. An internalist about testimony, such as Elizabeth Fricker (1987) or Jonathan Adler (2002), maintains that trust in the content of testimony is grounded in the recipient's drawing on (typically tacit) evidence about who qualifies as a trustworthy testifier. In that case, we still credit the information imparted because we credit the informant. The answer to "How do you know?" or "What makes you think so?" remains "She said so." The dichotomy between teaching and testimony remains.

Well-designed exams test for what educators think the students ought to have mastered. Criteria for assessing exams align with the designers' desiderata. These points are uncontroversial. So requiring students to justify, explain, or show their work indicates that the test designer does not hold that the goal of the course (or unit or lesson or program) is primarily or exclusively the acquisition of particular bits of information. And the refusal to accept "the teacher said so" as an answer indicates that the goal is not primarily or exclusively the acquisition of bits of knowledge. Rather, in asking students to explain, justify, or show their work, we indicate that our goal is for them to learn to think with and about the topic in ways that reflect the methods, goals, and standards of the discipline. Dewey holds that bits of knowledge are resources to think with (1916:152). If so, it explains why philosophy professors do not treat students' disagreement with what we said in our lectures as evidence that we failed. If we have taught our students to think well enough that they can mount a serious, well-framed, well-supported defense of their conclusions, we have achieved an important pedagogical objective whether they agree with us or not.

Obviously, the reasons students can adduce vary with subject and grade level. When asked about the origin of the human species, we would not expect Stella's fourth graders to display the same sophistication as graduate students in evolutionary biology. But we would expect even fourth graders to adduce reasons bearing on both evolution in general and on the relation of *Homo erectus* to *Homo sapiens*. To be sure, an answer to such a question (even if not one expected of a fourth grader) might discuss how experts know. Nevertheless, a satisfactory answer would not consist of simply deferring to an expert.

Teaching for Understanding

Explain, describe, illustrate, show your work—such instructions frame homework assignments, problem sets, term paper instructions, and exams. Students are expected to display their grasp of the topics the assignments concern. To succeed, it is typically not enough for a student to parrot back the information she has been given or robotically apply rules she has been taught. Nor is it enough that she come to believe it and believe that it is warranted. The recipient of testimony does both. Unlike the recipient of testimony, the successful student is supposed to accept that p in a way that equips and entitles her to use it as a basis for further reasoning about the subject matter. She should be able to do more than parrot or paraphrase what she has been told, more than draw obvious inferences, and more than follow the exact instructions imparted. She should be able to extrapolate. This might involve identifying other cases of the same or similar phenomena or explaining why this phenomenon is unique. It might consist in projecting into the past or the future, beyond what was explicitly covered in class. It

Beyond the Information Given

might be a matter of being able to identify and accommodate a suitable range of counterfactuals (see Hills 2016; Grimm 2006) or to draw fruitful analogies (see Nersessian 2008). Stella's students can do these things. They can explain why members of *Homo erectus* who had certain traits had an advantage over their conspecifics. They can explain why, if the next generation inherited those traits, they would share the advantage the traits conferred and why, over many generations, their descendants would come to predominate. They can infer that the same process occurred in the evolution of modern-day giraffes and squirrels. They can fruitfully speculate that had the environment been somewhat different, modern humans (or giraffes) would have evolved to have different traits. Such responses are well within the abilities of fourth-grade students.

Eventually, students should learn to critically evaluate the material they are taught. They should learn to assess the strength of the evidence for the positions they come to hold (see Siegel 1988:32–47). To accept p is to be willing and able to use p in nontrivial inferences and actions when one's ends are cognitive. p is acceptable only if accepting p is appropriate—that is, if one's reasons for p satisfy the standards reflectively endorsed by the relevant realm of epistemic ends. In that case the epistemic agent has reasons that the members of the relevant realm of ends ought not reject, for both the verdict and the reasons for it satisfy the community's standards (see Scanlon 1998:32–33). But an agent ought not accept p if she is unaware of its acceptability. To be a responsible acceptor of p, she needs to recognize (or at least have good reason to think) that it is acceptable. Hence, she needs to (1) recognize her reasons, (2) recognize that they are reasons for p, and (3) appreciate the force and weight of those reasons. This enables her to recognize that her reasons meet or surpass the community's (and her own) threshold for acceptability. Exercising epistemic agency requires critically assessing reasons. Thus educating students for epistemic agency requires teaching and motivating students to reason critically.

Greco takes understanding to be systematic knowledge and takes lecturing in fact-focused courses to consist in testifying to the effect that the systematic truth obtains. On this view, testimonial uptake would equip students to recognize and perhaps extract elements of the system. If Timmy knowledgeably testified that a & b, the recipients of his testimony would be in a position to know that a. But such knowledge does not equip the recipients to go much beyond the information given. It would presumably enable them to make trivial inferences. But they would not be equipped to make substantive, ampliative inferences or to draw fruitful analogies and disanalogies. A fundamental goal of teaching is to enable students to reason beyond what they have been explicitly told and beyond what is trivially obvious on the basis of what they have been told. They should be able to use the facts they learn, as well as the methods, orientations, standards, and approaches to stretch the limits of their thinking in epistemically fruitful ways. They should be able to leverage their understanding to gain further understanding. Successful uptake then is more than registering and crediting the information imparted. It involves developing a capacity to exploit that information to engender further understanding.

Disciplinary Understanding

From the earliest grades, students study history, a discipline that seems thoroughly grounded in facts. We might suppose that the goal is for them to come to know important historical facts—for example, when the Battle of Hastings occurred or who invented the steam engine. Standard instructional practice belies this. History teachers are not satisfied with the performance of students who merely know what happened. To see why they are not, let us distinguish between a chronicle and a historical narrative (see White 1965:222–225). A chronicle is a record of facts about the past; a historical narrative establishes connections among them. The distinction is conceptual, not chronological. Although a chronicle provides data for a historical narrative, we should not imagine that its chronicle is complete before a historical narrative is written. A chronicle and the associated narrative each influence the development of the other. As a history emerges, a historian realizes that the chronicle needs additional facts. Still, the distinction is epistemologically useful in that it enables us to isolate different elements in our understanding of the past.

A chronicle is just a list of facts. It makes no connections. The position of an entry on the list is arbitrary. No order is even implicit. The individual entries are items that could be imparted by testimony. So, for example, a chronicle of facts about Julius Caesar available to fourth graders might include:

Died: 44 BCE

Roman general

Born: 100 BCE

Beyond the Information Given

Killed on the Ides of March

Kidnapped by pirates

First Roman emperor

Crossed the Rubicon

Fought in the Gallic Wars

Married three times

Marched with his army to Rome

If the instructional objective were simply that the students know these facts, the teacher might just require them to memorize the chronicle. Successful students would then reel off the facts by rote. But the history teacher's objective is different. She wants the students to begin to understand Caesar's rise and fall, the ways it sprang from earlier events and set the stage for later ones, the ways it impacted the history of the Roman Empire and of the West. That requires a history, not just a chronicle. The students should learn to appreciate how the facts listed in the chronicle relate to one another as well as to further matters that do not appear on the list.

A historical narrative organizes facts listed in the chronicle, relating them to one another. It establishes temporal order, causal relations, dependencies; it makes logical connections, draws distinctions, provides explanations. It uses words like "because", "in order to" ,"after", and "therefore", which are not to be found in the chronicle. The historical narrative omits and augments. For example, an elementary school–level history of Caesar's rise and fall might omit mention of his multiple marriages on the ground that they don't seem to matter to the understanding it seeks to provide. It might augment "crossed the Rubicon" to emphasize that Caesar crossed from the north to the south because he was heading for Rome. It might contend that his intention to lead his army to Rome explains his crossing the Rubicon. It might take his crossing the Rubicon as evidence that Caesar was ambitious. It might go on to suggest that his ambition led to his assassination. The history then weaves the facts of the chronicle into a narrative that makes sense of the episode it deals with.

Although the story seems simple, the narrative is epistemically complex. To convert a chronicle into a history requires criteria of relevance, evidence, and importance. Decisions about ordering, augmentation, and omission go beyond the facts that the chronicle supplies. Taxonomy and vocabulary may be crucial. Does the available evidence support the contention that Caesar was ambitious? Does it support the contention that he crossed the Rubicon because he was ambitious? Answers to such questions depend on the criteria of acceptability in play. They determine whether the chronicle supplies the sort of evidence required to attribute character traits and motives. Through the historical narrative, the students begin to glean insight into such matters.

The students are given the narrative, not just the chronicle. They may experience it as a seamless whole, telling the story of Caesar's rise and fall. It might seem then that all the epistemological work is done by the writer of the text; the students are oblivious to it. But the seemingly seamless narrative admits of a sort of epistemological factor analysis—a factor analysis that figures in what the students are expected to do with the narrative. They need to take it apart. They may be asked to distinguish between the brute facts, which would appear in the chronicle, and the interpretive elements, which would figure in the explanations of those facts. For example, they may be expected to recognize that the sorts of considerations that could reasonably be adduced to argue that Caesar was ambitious are different from the sorts of considerations that could reasonably be adduced to argue that he fought in the Gallic Wars. They may be expected to distinguish between important and unimportant facts. Does it matter that he was kidnapped by pirates, or is that just an odd bit of trivia? In preparing to write an essay, they may be advised to start by making a chronicle of the facts they want to include, then to go on to write an account that connects those facts.

In describing a chronicle, I said nothing about what qualifies a statement of fact for inclusion. Even though a chronicle is just a list, it is not an arbitrary list. It is a list of facts about a particular episode, event, or era. We would not find "Platypuses are monotremes" or "The Red Sox won the 2004 World Series" in the chronicle we have been discussing. They have nothing to do with Caesar. Nor would we find "Caesar disliked beans," since even if it is true, there is no evidence for it. Evidently, statements of fact qualify for inclusion in a particular chronicle by being recognized as satisfying disciplinary demands for accuracy, relevance, and justification. The discipline of history underwrites the statement that Caesar was killed on the Ides of March. It certifies that the statement satisfies its standards. Disciplinary norms thus figure in establishing criteria for inclusion in a chronicle. Omissions matter. If the chronicle omits important, available information, or the narrative excludes or elides it, the history is flawed. So the student who begins work on her essay by writing a chronicle should recognize that the facts that she lists

Beyond the Information Given 227

should be ones that historians would deem relevant, accurate, sufficiently well established, and important, and that the ones she omits are irrelevant, unimportant, or untrustworthy. To satisfy that requirement, she needs at least an implicit grasp of the discipline's criteria.

The narratives, both those the students read and those they write, may disclose gaps and incongruities. Questions arise, grounded in the connections that have been forged. If Caesar was a general, fighting a land war in Gaul, how did he even encounter pirates, much less get kidnapped by them? How did his participation in the Gallic wars bear on what happened when he moved on Rome? As her understanding of history grows, the student should be able to identify significant gaps, incongruities, and biases in the emerging account and begin to recognize or develop strategies for resolving them. Minimally, she should appreciate that the gaps, incongruities, and biases show how and where her current understanding is limited. For this, she needs to take a critical stance. Even a fourth-grader's understanding of Caesar's rise and fall involves considerably more than knowledge of discrete facts. Making sense of the episode requires respecting the relevant epistemic standards, norms, and criteria. It involves making the sorts of connections that satisfy the grade-appropriate standards of the discipline and eschewing those that do not.

A similar schema applies to the emergence of understanding in other disciplines. Although White applies his conception of a chronicle exclusively to the discipline of history, we can think of a student's scientific understanding as based on a chronicle of scientific facts—perhaps facts about covalent bonds. The chronicle might consist of a list of covalent compounds:

Oxygen—O_2
Chlorine—Cl_2
Water—H_2O
etc.

It might also include statements like "Covalent compounds share two or more electrons. Again, these could be conveyed by testimony.

The scientific chronicle would not include the statement that the compounds are covalent *because* they share two or more electrons. It would not say how they come to share electrons or why it matters that they do. Causal and explanatory connections go beyond the material expressly presented in the corresponding chronicle. A scientific account, like a historical narrative, systematizes and organizes the material in its chronicle to establish logical, spatiotemporal, and explanatory connections. Unlike many historical narratives, however, a scientific account typically includes models that serve as mediators (see Morgan & Morrison 1999), linking individual matters of fact with overarching scientific laws. We understand the facts by, in effect, filtering them through the mesh that a model provides. The models are not themselves statements of *fact*, however, for they are known not to be true. I characterize them as felicitous falsehoods (Elgin 2017). Others take them to be approximations or quasi-truths (see Khalifa 2017; Grimm 2016). Either way, an understanding that represents covalent bonds as Lewis structures, or one that appeals to $pV=nRT$ to explain the relation between temperature, pressure, and volume in a gas, does not restrict itself to literal truths. The permissibility of such deviations derives from the science's conception of the sort of understanding it seeks to provide. That conception underwrites the conviction that the deviations from truth are, in the context where the models function, not difference makers (see Strevens 2008:55–56). A student incorporating such models and laws in her understanding of the phenomena ought not, of course, take them to be true. Rather she needs to appreciate both that they are not strictly true and that their divergence from strict, literal truth does not discredit them.

To go from a scientific chronicle to a systematic understanding involves establishing relations that underwrite explanations, observations, demonstrations, and experiments. As in the move from a chronicle to a history, some elements of the scientific chronicle may be set aside on the grounds that they are mere curiosities or outliers that the science need not accommodate, or on the grounds that they fall within the province of a different discipline. Factors that were not listed in the chronicle may be introduced. These might include additional covalent compounds, distinctions between types of covalent bonds, intermediate steps that need to be filled in, as well as new or refined models and idealizations. They are justified to the extent that they strengthen the epistemic network.

The emerging account must satisfy appropriate criteria of evidence and relevance. It must exclude considerations that are deemed scientifically impermissible. Although it may be reasonable for the historian to adduce Caesar's ambition to explain his crossing the Rubicon, it would be impermissible (except perhaps metaphorically) for the scientist to adduce the atom's ambition to complete its electron shell to explain a covalent bond. Like the

Beyond the Information Given

student of history, the student of a science must be sensitive to the gaps and incongruities in her nascent understanding. She should recognize questions it leaves open and should have some idea how to approach them scientifically. She needs to understand the relevant scientific methods, what they deliver, and why and to what extent their results are creditable. Here, too, understanding goes beyond knowledge of established facts. The successful chemistry student must do more than memorize the chronicle. She needs to grasp the connections the science finds among the items listed in the chronicle. She also needs to appreciate why the science takes these connections to hold, to be explanatory, and to be significant. Only with an appreciation of the relevant methods, norms, and standards does the student understand the subject matter.

This way of putting things may sound intellectually too sophisticated to characterize K–12 student learning. But it is, I suggest, what students achieve when they come to understand a topic. That understanding dawns slowly. There is no suggestion that young students are self-consciously aware of the norms, standards, and criteria implicit in their substantive grasp of a subject matter. But even young students are regularly asked to explain, to give examples, to extrapolate to further cases, to draw inferences that go beyond the explicit content of the instruction they have been given. To do so, they need to be at least implicitly aware of the epistemic norms, standards, and criteria that govern the discipline. Over time, I suggest, if they continue in the discipline, what was implicit becomes explicit. They learn how to think like a historian or a chemist or a geographer. As they internalize and endorse the epistemic norms and standards, those norms and standards provide a basis for critical assessment of the ways the discipline approaches its subject. Although this may not be explicit, they come to understand not just the subject but the nature of their understanding of the subject.

Extrapolating

Students who have mastered the material are expected to be able to go beyond the information given. Trivial inferences count for little. They simply articulate what was obvious anyway. But students should be able to solve further problems, project to additional cases, draw effective analogies, generate plausible hypotheses on the basis of what they have learned. This is the educational dimension of the requirement that the network enable nontrivial inferences and actions. Its importance is not exhausted by what it shows about current mastery. A network in reflective equilibrium should leverage further inquiry. It should equip students to build on what they currently understand. In demonstrating their mastery, the students discover that they can do more with the material than what they have been explicitly taught. This puts them in a position to see not just that but also how the subject is open-ended. There is more to be discovered.

The growth of understanding is flexible, fallible, and dynamic. A network's equilibrium may be upset by new findings—findings that its methods enabled it to uncover. This is an asset, not a liability. It enables us to remove previously accepted errors, ill-advised strategies, unreliable methods. Whether or not the current equilibrium is flawed, it is open to further elaboration and expansion. So the acceptability of a given network of commitments is not expected to be permanent. New questions, techniques, and standards are apt to raise doubts or disclose limitations. In light of new considerations, the network is susceptible to reevaluation. That being so, there is benefit in students revisiting material they previously studied. More is involved than learning a few further facts. When students who studied Caesar in fourth grade learn more about the history of Rome in later grades, they may come to suspect that Caesar's multiple marriages, which they had dismissed as merely personal, actually played a role in forging important political alliances. They then have reason to integrate Caesar's marriages into their emerging understanding of the period. That gives them an incentive to reconsider other considerations that they had set aside. Because the networks of epistemic commitments that constitute understandings of a topic do not purport to provide the last word, they are springboards for the advancement of understanding. The recognition that current understanding is limited and may be flawed makes sense of how investigators proceed at the cutting edge of inquiry. It also makes sense of what goes on in education. There is then a continuum from the earliest education up to and beyond the cutting edge.

A good exam asks students to manifest their disciplinary understanding of the material being studied. A high school chemistry student taking an exam on covalent bonding will cast her answers in the language of chemistry: molecules, atoms, and electrons, as well as Lewis structures, orbitals, and bonds. That terminology marks the distinctions that are deemed to be important to chemistry's understanding of its domain. In properly using that terminology, the student shows a recognition of how the science frames its topics. Although she will make some literal statements of fact—for example, "H_2O is a covalent compound"—her answers are apt to involve statements and diagrams that describe the phenomena via models and idealizations that are not literally true. In relying on models and diagrams, she is no different from professional chemists. To be sure, theirs are more sophisticated. But both the student and the professional understand covalent bonds in terms of models that diverge from the phenomena. For her answer to be duly responsive to evidence, she must draw on and frame it in terms of the sorts of considerations that chemistry counts as evidence. It will not do to simply assert that she has it on good authority that H_2O is a covalent compound, even though it is, and the expert she relies on is epistemically responsible. Her answer should, at least implicitly, reflect that she recognizes what evidence is relevant and why. She should make it clear that it is reasonable to reflectively endorse her conclusion , given evidence of this kind. Her understanding of covalent bonds should enable her to draw nontrivial inferences about the subject. She thus needs to be sensitive to the kinds of inferences that are acceptable in (high school) chemistry and which of these are relevant to the question. She needs to be aware of what considerations and modes of argument she can draw on to back her claim. She should recognize what factors she cannot afford to omit. The student is expected to be, and to display that she is, attuned to the methods, standards, resources, and orientations of the discipline. It is not enough to give a list of known facts about covalent bonds. Depending on the question, she may be asked to go further—not only to draw nontrivial inferences but perhaps to speculate about what would be the case if a bond were weaker.

Reflective Endorsement

An understanding of a topic is a network of commitments in reflective equilibrium. *To* understand a topic is to accept such a network—that is, to be willing and able to use it in nontrivial inferences and actions when one's ends are cognitive (see Elgin 2017). A willing but clueless chemistry student has no grasp of covalent bonds. An unwilling but clued-in chemistry student is loath to reason and act on what the network of commitments provides. The ability requirement involves competence. For Mariah to understand the phenomenon, she must be able to reason appropriately about, and engage in appropriate actions regarding, covalent bonds. This might involve inferences, analogies, extrapolations. It might involve designing and executing experiments or contriving idealized models. It does involve appreciating the limits within which her reasoning and actions are appropriate. If she is able, she can do such things. Willingness is a manifestation of reflective endorsement. In being willing to accept the network, Mariah adopts it as her own. She acknowledges that it provides resources to promote her cognitive ends with respect to a branch of chemistry.

Epistemic acceptance is a higher-order stance; it involves more than merely appreciating what the network is committed to. Lackey's Stella shows this. Stella is adept at reasoning within the theory's network of commitments. She can show how and why that network supports the claim that *Homo sapiens* evolved from *Homo erectus*. She readily works out the implications of the theory. But she does not reflectively endorse the theory or its implications. She does not appreciate the force of the reasons she adduces. She considers the theory and its implications wildly incorrect. She recognizes both that and why the theory of evolution is the consensus opinion in biology. But she does not understand the origin of species in terms of it, for when her ends are cognitive, as opposed to merely pedagogical, she is unwilling to use it. She refuses to stand behind it. Her students take the further step. Convinced by her teaching, they hold that the theory of evolution accounts for the diversity of life on Earth. They reflectively endorse the theory. They are willing and able to draw on the resources it supplies to provide reasons for biological claims.[3] Unlike their teacher, they understand the origin of species.

To reflectively endorse a network of commitments is to take epistemic responsibility for it. The agent must consider herself willing and able to supply reasons in defense of the network and the inferences and actions it licenses. She thus has to recognize the relevant reasons and their strength. Again this may sound epistemically too ambitious. But it amounts to her being in a position to give a cogent answer to questions like "Why should we think *that*?" If a fourth grader gives a grade-level appropriate answer to such a question, she discharges her epistemic responsibility. Recall that a network that provides an understanding equips those who reflectively endorse it with the capacity to make *nontrivial* inferences. Such inferences are neither easy nor obvious. So a student who understands a subject both needs to, and is equipped to, use her own judgment about the subject. Although what she has been taught supplies resources, she has to go further—to manipulate the commitments to answer new questions, formulate and solve new problems, perhaps set new epistemic ends. This is an exercise of epistemic autonomy.

Beyond the Information Given

A worry arises. Inasmuch as the networks a student endorses are largely the fruits of her education, might her pretensions to epistemic autonomy be spurious? If she considers *e* to be evidence for *p* only because this is what chemistry counts as evidence, or if she considers *s* to be a reason for *q* only because that is what history counts as a reason, she seems to be a victim of indoctrination. Her opinions are not her own. That she stands behind the disciplinary mandates shows only that the indoctrination was effective. But education is not indoctrination (see Callan & Arena 2009). It involves both the development of both critical thinking skills and the propensity and motivation to deploy them (see Siegel 1988:78–90). If the student both appreciates that this is what chemistry counts as evidence and is convinced that chemistry gives her good reason to count it as evidence, she reflectively endorses the standards by which chemistry judges such things. She is a budding member of chemistry's realm of epistemic ends.

Students should learn both why practitioners in the different disciplines favor their criteria of reason or evidence and how those criteria can responsibly be challenged. This is not so difficult as it might seem. Even young students have the ability to recognize when a consideration they are inclined to credit does not fit comfortably with what they have been taught. When a student who has been taught that every event has a cause learns that radioactive decay is stochastic, she recognizes that the new information does not fit with what she has learned. When a student who has been taught that justice is blind learns that a disproportionate number of African Americans are convicted of crimes, she has reason to question what she has been taught. Even if she is in no position to resolve the tension, the recognition itself gives her a stance for thinking critically about what she has learned. This is not to suggest that she always will or always should abandon what she has been taught. But the tension she recognizes is evidence that all is not epistemically well. If the student has been deprived of cognitive resources for looking further or has been disincentivized from looking further, she is a victim of indoctrination. If she recognizes the mismatch, she can develop the incentive to look further and has some idea about how to go about looking further. She is then capable of functioning as an epistemically autonomous agent. If she is motivated to look for mismatches and investigate what they reveal, her education has served her well.

Equilibrium can be destabilized by new insights, new information, new methods, new perspectives. So rather than treating what they have learned

234   Chapter 9

so far as fixed and final, students need to learn when and how to revise their views. Rarely, if ever, are there clear-cut decision procedures for making such revisions. In learning to think critically, students develop the skills, motivation, and propensity to consider what is to be said for and what is to be said against accepting (or continuing to accept) a commitment c; what epistemic and non-epistemic risks accompany accepting c, rejecting c, suspending judgment about whether c; what plausible alternatives there are to c, and what is to be said for and against each of them. In acquiring these skills and propensities, they learn to use their own judgment. To use one's own judgment is not a matter of deciding on the basis of a whim or a personal preference. It is a matter of weighing alternatives when the answer is not clear (and sometimes when grounds on which to weigh are themselves matters of controversy). Since whatever an epistemic agent chooses is what she reflectively endorses, she recognizes that she is expected to be able to defend her choice, to provide reasons (even if not conclusive reasons), to stand behind her choice and behind her reasons for making that choice.

Understanding

Although I have spoken about fact-focused disciplines and pointed to a couple of examples, I have not said what makes a discipline fact focused. Aren't all academic courses fact focused? Math courses concern facts about mathematical entities—numbers, sets, functions, and the like. Literature courses concern facts about novels, poems, stories, and the like. Philosophy courses concern facts about knowledge, goodness, reality, and the like. The issue here is not whether the disciplines concern bodies of facts. All serious academic disciplines do. The issue is whether lessons in those disciplines mainly seek to impart those facts. On Greco's view, teaching history or science turns out to be fundamentally different from teaching math or poetry. Even if delivered in lecture form, a lesson on calculating the area under a curve or on closely reading a Shakespearian sonnet is hard to construe as aiming to impart truths about matters of fact. The area under that particular curve is not very important. Rather, the point of the lesson is that the strategy used to calculate this area works for a wider range of cases, and that its working for that range reveals something significant about a class of curves or a method for calculating areas. Similarly for analyzing a poem. There is, perhaps, some value in focusing on weather metaphors in

Shakespeare's *Sonnet 18* for their own sake, but the student is expected to learn more than how these particular metaphors contribute to the meaning of this particular sonnet. She is supposed to learn how to interpret images in Elizabethan poetry and how imagery generally functions in literature, perhaps even how to see the world through the lens that the poem provides. But, Greco seems to think, the natural and social sciences are different. Their main goal is to impart facts.

I disagree. Little teaching in any academic discipline is primarily concerned with imparting facts. If fact-focused courses are courses whose main epistemic objective is to impart facts, there are few if any such courses; if they are courses where the understanding they seek to engender is grounded in facts, all academic courses are fact focused. Understanding a topic, I have argued, consists in grasping and reflectively endorsing a comprehensive, systematically linked body of information in reflective equilibrium where that body of information is grounded in fact, is duly responsive to evidence, and enables nontrivial inference and perhaps action regarding the phenomena the information pertains to (Elgin 2017:44). Education, I maintain, aims to engender understanding. This is so from elementary school through graduate school. Let's look at the several conditions.

Understanding should be *grounded in fact*. Elementary education is simplified and schematic. So are many introductory courses in universities. Insofar as the material is simplified, much of it is strictly false. Students learn Newtonian mechanics as a stepping stone to relativity theory and quantum mechanics. Although Newton's laws are literally false, they are a valuable way station to a more accurate understanding of physics. The understanding the students gain is grounded in fact, insofar as it is true enough (see Elgin 2017). Middle school science students gain some understanding of the relation of force to mass via $f=ma$, even though the formula is strictly false because it fails to accommodate relativistic effects. Students gain an understanding of a subject, insofar as the divergence of the information they learn from the facts it pertains to is negligible. Eighth-grade physics students and fourth-grade biology students have grade-level-appropriate understandings of their subjects because the relevant physical and biological facts ground that understanding. Those facts figure crucially in why the schematic, elementary explanations work as well as they do.

The simplifications taught in the early years are not chosen merely because they make things easier. Rather, they provide a basis for leveraging understanding. $pV=nRT$, although strictly false, enables students to begin to understand thermodynamics. Students can grasp the importance of the interdependence of pressure, temperature, and volume in a gas. This puts them in a position to realize that and to realize how and why things are more complicated and how and when and why the additional complications matter: gas molecules collide with and bounce off one another, they are subject to gravity, they have a variety of shapes and sizes that affect what happens when they collide, and so forth. A good course not only provides age-appropriate information, it positions and equips students to take the next steps.

Perhaps some understanding is so firmly grounded in particular facts that it is necessary to know (or justifiably believe) those facts in order to understand the subject at all. Given the reliance on simplifications in early education, I doubt that this is so. But it might be. Maybe one could not begin to understand the voyages of discovery if one did not believe that Earth is spherical. If there are such facts, then in order to understand the topic, any student of the subject must be privy to them. A teacher could simply state these facts and warn her students that they won't be able to understand the topic if they do not accept them. If the teacher simply presents herself as an expert and imparts such a fact, then at least at the outset, the students glean it as testimonial knowledge. If so, testimony has a legitimate place in understanding. In effect, the teacher is saying, "Here's a fact that your understanding of the subject must be responsive to. Take my word for it." But as a student's understanding develops, the testimonial status of the knowledge is supplanted by a more holistic justification. The student comes to see how the fact plays a crucial role in a more comprehensive understanding of the topic. And her justification for that judgment of fact becomes dispersed across her understanding of the topic as a whole. Her reasons for believing it no longer rest on the mere fact that her teacher told her so.

An understanding must be *duly responsive to evidence or reasons*. Students not only glean factual information about, as it might be, the evolution of species or the Norman conquest they also learn what counts as evidence for that information. That is, they learn about the importance of the fossil record in establishing well-grounded views about lines of descent or about the primary source documents and historical artifacts that afford evidence about the Battle of Hastings. They learn that without such evidence, claims about such topics are unfounded. Eventually, if not in the early grades, they learn

Beyond the Information Given

why certain disciplines require certain sorts of evidence, why some seemingly relevant considerations do not qualify as evidence, and how investigators gather and assess evidence. In nonempirical fields, such as mathematics and poetry, they learn to identify and assess reasons for and against a claim. The critical point is that students learn what justifies the considerations that the various disciplines endorse. They learn what qualifies as an acceptable answer to the question: Why should we accept *this*? This is one place where a testimonial model is inadequate. When the answer is "Because Professor Nolan said so," the reason—although reliable—is not suitably connected to the facts. If the topic is the Norman Conquest, the student's reasons for her belief should be grounded in facts about the Norman Conquest. The reasons must be topic related, not merely source related.

Here too, leveraging is important. Elementary school–level standards of evidence provide a basis for refinements. Students who initially thought they just needed evidence for their claims learn that evidence can be misleading. So they learn how to avoid being misled. They learn, for example, that an experimental result must be reproducible, and they learn what counts as reproducing it. They learn that to serve as evidence for a historical claim, primary sources are mandatory, and they learn that a single primary source document should be backed by additional evidence. And so forth.

An understanding *enables non-trivial inference and perhaps action* about the phenomena it bears on. The understanding should enable a student to draw responsible inferences beyond the obvious. This may involve extrapolating to the past or the future. It may involve counterfactual reasoning. It may involve drawing analogies or disanalogies to seemingly related cases. The inferences in question need not be formal. Often extrapolating beyond the information given takes the form of associative reasoning—projecting one's insights onto similar (or not so similar) cases. The inferences may also show why something that holds in a given case does not carry over to seemingly similar cases. Thus, for example, chemistry students should be able to explain why, despite the fact that most materials contract when they solidify, water expands. That is, they should be able to explain why water is unusual.

Leveraging here is a matter of increasing sensitivity to the scope and power of the inferences drawn. Stella's students might be able draw on what they learn about human evolution to speculate about the evolution of other organisms. If they have also learned about the importance of environmental pressures, they could be able to venture plausible hypotheses about why the evolution of parrots and the evolution of penguins took different paths. As their biological understanding increases, they come to realize that some inferences are limited in scope. Some of the things they learn about one organism can be expected to hold of all other normal healthy organisms; others can be expected to hold only of plants or only of members of a particular species, and so forth. Moreover, they learn that their counterfactual judgments need to be suitably circumscribed as well. If the climate had been a bit colder or a bit warmer, then the species would have or might have displayed p. If it had been drastically hotter or colder, perhaps there is no saying whether the species would have survived or what adaptations might have occurred. The more students understand about a range of phenomena, the more acute, refined, and sensitive their inferences are expected to be.

The commitments that constitute an understanding form a network in reflective equilibrium. They are reasonable in light of one another and as reasonable as any available alternative given the relevant antecedent commitments. This latter constraint must be keyed to an appropriate epistemic community. The community of fourth-grade biology students has antecedent commitments that they draw on in developing an understanding of evolution. It is looser and less reliable than the commitments of professional biologists. Nevertheless, by drawing on what the students know or can observe about biological inheritance—for example, that members of a family are apt to look alike—they can begin to get a handle on heritability. Not all of the original commitments will be retained in the system that emerges. Like other epistemic agents, students correct and refine their untutored beliefs. But they typically start with such beliefs. The system provides both resources and constraints. Coherence conditions both limit the inferences that can be drawn and provide opportunities for elaboration and expansion of a network of commitments. If this is so for humans, the students think, something similar should hold for other animals. Maybe not all animals, they then might think, but anyway for mammals; maybe not for all mammals, but anyway for all terrestrial mammals. We can see how such reasoning can extend and deepen their understanding. One pedagogical goal is to orient students to significant features and direct their attention to potentially useful lines of inquiry. Although entertaining counterfactuals is often fruitful, outside of philosophy, it is usually ill-advised to worry about very distant possible worlds.

I said that the network of commitments that constitutes an understanding should be reflectively endorsed. That requires that not just the students

Beyond the Information Given

accept the commitments but also that they recognize that the commitments stand up to scrutiny. To do that, the students must both be capable of and inclined to engage in critical reasoning (see Siegel 1988:33–47). They need to be responsive to reasons—that is, to be both able and motivated to recognize reasons for and against a claim; to assess those reasons; and to assign them due weight. They also need to be motivated and equipped to provide reasons for contentions that they think are worthy of reflective endorsement. That is, they need not only to be able to ask, "Why should we believe *this*?" They should also be equipped and motivated to recognize and accept good answers and recognize and reject bad answers. To reflectively endorse a consideration is to recognize that it satisfies the standards that ought to be satisfied for considerations of its kind. What the standards are, and how demanding they are, changes as one matures and becomes more sophisticated in a discipline. Here, too, the light dawns gradually. But even very young children ask "Why?" And quite early, they can be brought to recognize both that the question is legitimate and that there are good and bad ways of answering it (see Harris 2012). Of course, merely asking "Why?" is not enough. One major goal of education is to equip students to learn how to ask "Why?" effectively—that is, how to frame a question, approach a problem, draw on available resources, and marshal evidence so that an acceptable answer emerges. Stella and Lilith were successful because they did not tell their students what to think. They showed them how to think, and they did so in a way that made manifest something of the power of thinking about the phenomena that way.

Conclusion

Lackey's creationist teacher and Malfatti's climate science professor do not testify; they teach. Since teaching is not testimony, their seemingly surprising success does not bear on the dispute between reliabilists and evidentialists about testimonial knowledge. Evidentialists and reliabilists agree that although the testifier assures her audience that the information conveyed is suitably justified, she does not articulate the justification. Hence, it is entirely acceptable for her audience to be clueless about the evidence for a fact they come to know. They believe it and are justified in believing it because they have it from a source whose word they justifiably take. This, as I said, is valuable. It underwrites the efficient transfer of information to epistemic agents who lack the resources or incentive to acquire or vet the evidence for themselves. Teaching, however, equips learners to go beyond the information given. Students acquire the ability to use what they are taught as a resource to extend, deepen, and critique their understanding of the subject matter.

The divergence of teaching from testimony suggests that the justificatory terrain of epistemology is more variegated than we might suppose. Perhaps some knowledge and understanding answers to evidentialist constraints, some to reliabilist constraints. But even if testimony answers to evidentialist constraints, those constraints are different from the ones that teaching answers to. This raises the question whether there are other modes of information transfer with distinct epistemic profiles, powers, and limitations.

10 Constructive Nominalism

Introduction

Earlier I described reflective equilibrium as a product of mutual adjustments. Considerations are modified to bring them into accord. Achieving reflective equilibrium involves making trade-offs and accommodating ourselves to both temporary and permanent cognitive limitations. I argued that rather than requiring acceptable theories to be true, they need only be true enough to serve our cognitive purposes. Even if I made a strong case that reflective equilibrium is a reasonable epistemic goal for finite, fallible epistemic agents, there remains a question: Why should we think that a system in reflective equilibrium answers to the facts? Epistemology seeks to account for our understanding of the way things actually are. How does the account I present do that? Or is it an argument that we cannot do that and must settle for something less?

The answer stems from the ecological model. An ecological niche is a product of mutual accommodation. Organisms both adapt to an environment and alter the environment to make it a more suitable ecological niche. The choice of the phenomena that epistemic agents seek to understand is a function of both the environment and the epistemic resources that epistemic agents bring to it. Epistemic agents adjust their interests to what is accessible, given their resources, and modify their resources as a result of the information they access.

We know from set theory that extensions are abundant. Barring paradoxical cases, any collection—no matter how motley its membership seems—constitutes an extension. The members of an extension are similar to one another by virtue of their membership in that extension. Most extensions are semantically unmarked. We have no reason to recognize them or the similarities among their members. But all of the extensions, marked or unmarked, are equally real. The critical issue is which ones are worth noticing. That is where the interests, abilities, and resources of epistemic agents come into play. What boundaries should we care about? What differences should make a difference? Which extensions constitute phenomena of interest depends on what our interests are. Electromagnetically, there is nothing special about wavelengths between 380 and 700 nanometers. What makes that range of wavelengths special is that it is the *visible* spectrum. We have very good reason to single it out, to subject it to investigations that we do not subject other ranges to—indeed to subject it to investigations that we could not subject other wavelengths to.

Epistemic agency then is involved in fixing focus—in selecting and marking out the collections of phenomena that are worthy of recognition. This does not sacrifice the objectivity of investigations or their results. It does, however, show that epistemic agency runs deeper into our understanding of the world than we might otherwise think. In this chapter, I argue that objectivity is consonant with the sort of epistemic activity I have been describing and that constructive nominalism supplies the metaphysical underpinnings. This marks another step away from the purely spectatorial perspective and the view that a viable epistemology provides a transparent window on what is there anyway.

Predicament

Naive realism may be the standard metaphysical default. Many people seem to assume, without much thought, that the world is the way it is regardless of what anyone thinks, and that sentences are true just in case they describe the way the mind-independent world is. Naive realism does not discriminate. It applies across the board. Earlier I suggested that this is what children are apt to think. So are unreflective adults. A little reflection reveals that it cannot be right. The truth values of some sentences pretty plainly depend on what some people think. Mental predicates are one source of difficulty. The truth value of the sentence "Someone believes that aardvarks snore" cannot be independent of what anyone thinks. If no one believes that aardvarks snore, then the sentence is false. Humility is also a problem. If someone thinks that he is humble, he is not. His being humble thus requires his failing to believe that he is. Then there is the problem of fashion. Manifestly I can

Constructive Nominalism

be wrong about whether an outfit is fashionable. So can anyone else. But it is highly implausible that miniskirts or baseball caps could be fashionable if no one thought they were. Predicates like "fashionable" seem to depend on what people think, even if not on what any particular person thinks. Some philosophers believe that mind dependence is characteristic of evaluative predicates in general. No more than fashionableness, would properties like goodness and beauty be independent of what people think. Nor is it clear what roles such properties could play in the causal structure of the world. So, such philosophers contend, sentences like "The *Mona Lisa* is beautiful" or "Giving to Oxfam is good" should not be construed as describing the way the mind-independent world is.

Evidently, a retreat from naive realism is necessary. The question is where we should stop retreating. Many draw the line at scientific realism. Conceding the points about fashion, ethics, aesthetics, and intentionality, scientific realists insist that natural science attempts to discover the way the mind-independent portion of the world is, and when it is successful, its success consists in it accurately characterizing that portion of the world. I will argue that *constructive nominalism* is more plausible.

Nominalism, on one construal, consists in the repudiation of universals— properties, relations, and kinds. It is a matter of saying, "There is no such thing as. . . ." It concedes that green things are alike in being green, but it denies that we need to posit a universal greenness to account for that. Standard arguments for nominalism attempt to show how we can do without such posits. My approach is different. It is an argument against metaphysical privilege (see McGowan 2003). I hold that there is no principled reason for recognizing that some collections are real kinds unified by metaphysically real properties, while others are arbitrary collections whose members need have nothing in common. If we posit a property of greenness to account for the similarity that green things share, I contend, we posit a property of grueness to account for the similarity that grue things share—and similarly for every other extension, no matter how motley it seems. We can, if we like, posit such properties. But even if we engage in profligately positing properties, we have to isolate the extensions that matter. To avoid being inundated by irrelevant extensions (with or without accompanying properties), we must pare things down. What the term "constructive" adds is that we contrive the predicates to answer to our evolving interests (see Hacking 1999). The nominalist, Hacking argues, denies that reality "is naturally and intrinsically sorted in a particular

244   Chapter 10

way, independent of how we think about it" (1983:108). We construct systems of kinds that answer to our evolving interests and promote our evolving goals. The sort of nominalism I advocate then is not a rejection of universals per se but a rejection of the idea that reality privileges any specific universals. All of the extensions are equally real. Either all are unified by universals or none are. The construction, moreover, is dynamic. The kinds we recognize evolve in tandem with our emerging interests, methods, and resources.[1]

Scientific theories go beyond the evidence they are based on. They may adduce substructures, superstructures, dispositions, causes, or grounds to explain what occurs. Their explanations thus yield entailments and implicatures about other actual and possible cases and about unexamined aspects of the phenomena they pertain to. Theories weave evidence into cognitive networks that organize data and enable us to leverage our understanding of a domain. This is why they are valuable. But what makes them valuable also makes them vulnerable. In going beyond the evidence, theories take risks. An explanation that accommodates all available evidence could be wrong. New evidence might turn up that indicates that, for example, estrogen prevents heart attacks after all. Even if no new evidence happens to turn up, a theory could get things wrong. Although we have no evidence of them, perhaps there are exceedingly rare extremophiles living at the bottom of the Mariana Trench. Still, we may think, our epistemic vulnerability is due to the dearth of evidence. Were it not for practical impediments, we could acquire enough evidence to render our theories invulnerable.

Let us grant that, ceteris paribus, more evidence is better than less. So, imagine a best-case scenario. Suppose that our theory accommodates all the evidence—past, present, and future—and that there are no practical impediments to gathering any evidence we want. Suppose further that our theory displays the full panoply of theoretical virtues. It is simple, predictive, fruitful, robust, and so on. Even then we have a problem, for another theory might be equally good in all these respects. Because theories are underdetermined by evidence, more than one theory can accommodate all the evidence, however plentiful the evidence is. Although the requirement that they satisfy theoretical constraints enables us to dismiss ad hoc or rococo alternatives, there is no reason to think that a unique solution results. Multiple empirically equivalent accounts are apt to display an equal balance of theoretical virtues (see van Fraassen 1980).

Constructive Nominalism

Such underdetermination should probably distress the realist. Very roughly, realism is the view that things are the way they are, regardless of what anyone believes about the way they are, and that the true theory is the one that describes the way things are. But it follows from underdetermination that we cannot determine how things are. At best, we might conclude that a disjunction is true: "*Theory A* or *Theory B* or *Theory C* or . . ." is true, but in principle, we will never be able to tell which one. This is not nothing, but it is disappointing. However good our particle theory, we will never be in a position to know, for example, that the world consists entirely of particles if an equally good alternative says that it consists entirely of waves. This does not refute realism. It simply underscores an unfortunate epistemological consequence. Realism leads to skepticism, not because of the inevitable paucity of evidence but because of the availability of alternative, equally good (although perhaps unformulated) theories.

The situation is this: *Theory A* (for example, a particle theory) is an ideal theory. It answers to all the evidence—indeed to all possible evidence—past, present, and future. It displays a maximal constellation of theoretical virtues as well. If *Theory A* were the only theory, we would accept it in a minute. It has everything going for it and nothing going against it—except for pesky *Theory B* (a wave theory), which is equally good in all relevant respects. If we conclude that we can accept the disjunction but cannot accept either disjunct, we concede that scientific inquiry, however successful, cannot yield knowledge of the way things are. At best it yields knowledge of a plurality of ways, one of which is the way things are. We cannot simply conclude that any particular disjunct is true, since we have no basis for rejecting the others. What should we do?

Bas van Fraassen thinks we should opt for constructive empiricism. Which theory is true is determinate. But it is something that we can never know. Nor does it matter. "Science aims to give us theories which are empirically adequate; and accepting a theory involves as belief only that it is empirically adequate" (1980:12). A theory's success then is a matter of empirical adequacy. There is no reason to think that science, even at the limit of inquiry, yields truth about anything beyond the observable phenomena, hence no reason to think that it yields knowledge about anything except the observable phenomena. This too is disappointing. We hoped for more than a sophisticated device for saving the phenomena.

Another alternative is this: rather than thinking that the adequacy of the theories gives us grounds for accepting the disjunction, or flipping a coin to decide between them, why not conclude that it gives us grounds for accepting the conjunction? Following Israel Scheffler (2008), let's call this view plurealism. If we have sufficiently good reason to believe that A is true and sufficiently good reason to believe that B is true, why not say that we have sufficiently good reason to believe that $A \& B$ is true?

Ockham's razor evidently cuts off this option. We ought not multiply entities beyond necessity. But to endorse the truth of $A \& B$ seems to do just that. If the world is comprised of particles, there is no need to admit waves into our ontology; if it is comprised of waves, there is no need to admit particles. Moreover, in a world comprised entirely of waves, there is no room for particles; in a world comprised of entirely particles, there is no room for waves. There is no logical inconsistency in the conjunction of A and B. But although each is independently tenable, the two are not cotenable.

Realism leads to skepticism; constructive empiricism settles for empirical adequacy; plurealism engenders severe overpopulation. All are disappointing if we seek through science (or any sort of systematic inquiry) to discover the way things are. But from the fact that a conclusion is disappointing, we should not infer that it is false. So perhaps our alternatives really are this bleak. Maybe even ideal science cannot disclose the way the world is.

Nevertheless, there is something weird about this predicament. Our problem is not a dearth of resources it is an embarrassment of riches. If we had only one of the theories, we would have no reservations about considering it true and concluding that the world consists of whatever it says the world consists of. After all, that theory satisfies all the relevant theoretical and evidential requirements we can devise. What better reason could we have for thinking it is true? What more could we want? The problem is that a second, equally good theory is too much of a good thing.

This suggests that we are misconceiving the problem in such a way that success looks like failure. Rather than thinking of truth as the *ur*-value of a good theory—that which makes a good theory good—perhaps we should think of truth as constituted *within* a good theory, where the goodness of the theory is determined by other factors. How does such a revision reconceptualize theories?

Realists, and many ordinary people, think that a scientific theory is, or should be, a mirror of nature. It should reflect without distortion the way

Constructive Nominalism

things actually are. Obviously, a theory must do more than that. To afford an understanding of a range of phenomena, it needs to find or make order, discriminate between relevant and irrelevant factors, point to some underlying or overarching regularities that the phenomena display, and so on. No mere mirror is so helpful. But however we spell out these details, "without distortion" remains crucial. Theories should, we may think, yield neutral descriptions of the way the phenomena are and would be, whether they were so described or not. It's a nice idea, and it explains why partiality and bias in theories are defects. But it may be too simple.

Perhaps rather than reflecting, theories refract. In that case, although no theory describes the phenomena as they are in themselves, each describes the phenomena as seen through a particular refractive lens. A characteristic of refraction is that the path of a light wave depends on the medium of refraction. This does not make the path less than objective. But it does mean that if we want to understand what happens to a light wave when it goes from one medium to another, we need to know the refractive indexes of the media. Similarly I suggest, rather than thinking that we can just look through a theory and see the mind-independent world, our encounters with things are always mediated and modulated by cognitive commitments. The result is an understanding of the-world-as-modulated-through-a-particular-theory. Truth then is defined within a theoretical framework rather than outside of it. But this does not make the truth of a sentence less than objective. It simply means that a claim to truth contains a (perhaps tacit) reference to the theoretical framework in terms of which truth is claimed. Given that we are in a theoretical environment that recognizes waves, the sentence "under condition c, wave form w occurs" is true, or it is false. The theoretical framework sets the conditions on truth; the phenomena either comply or do not comply. So the truth of p is independent of our belief that p is true. The world as described by the theory remains independent of our attitudes. But the kinds in terms of which we classify things, hence the sentences that are candidates for truth, depend on our cognitive constructs.

Metaphysical realists might argue that what I am saying is trivial. Of course, if we had never invented the word "rock," the sentence "New England's soil is full of rocks" would not be true. But since all we are doing in introducing a word like "rock" is making up a label for what is out there anyway, our contribution is negligible. What I have called "the-world-as-modulated-through-theory-T," they would maintain, is simply the world as described in the language in which *T* is expressed. Their picture is this: a scientist looks into her microscope, sees a cellular mass, and mutters to herself, "Lo! A hitherto unobserved object! I need a word for it. I will call it a 'mitochondrion.'" In that case, the label is just a tag for something that was there to be seen. The significant thing she has done is detect a constituent of the cell, which was there all along, even before she or anyone else noticed it. But this makes things too simple. Individuation and classification involve more than pasting on labels. To individuate a material object is to demarcate its spatial and temporal boundaries and to specify the changes it can undergo while remaining the same thing. To individuate a property is to specify the variations among instances that are compatible with their instantiating the same property. To classify objects is to set criteria for belonging to the same kind. How can items of this kind differ from one another and still count as the same kind of thing? Which things count as mitochondria and why?

A hardcore realist might maintain that every difference makes a difference to the identity of an object. The world does not discriminate between important and unimportant properties. So if the world determines identity, every difference between objects makes them different sorts of things, and every change an object undergoes makes it a different thing. Such a metaphysics might accurately reflect the way the world is, but it would be utterly unwieldy. We could not fathom it. We could not understand in terms of it.

Privilege

To determine what there is, we need to settle which differences matter, thereby marking out individuals, properties, and kinds. The scientist who discovers mitochondria does more than notice and name a blob on her slide. She explicitly or implicitly identifies that blob with some intracellular bodies and differentiates it from others. Of course this is not typically done all at once, and need not be done by a single scientist. Maybe someone does look into a microscope, notice a previously overlooked blob, and give it a name. But further work needs to be done. The blob has to be differentiated from other intracellular bodies. It has to be found in other cells. It has to be characterized in such a way that scientists can tell whether something is a mitochondrion; can tell whether, having undergone various changes, it is the same one; can tell when it ceases to exist as a mitochondrion; and so forth. This is not a matter of merely coining a term or labeling an antecedently

Constructive Nominalism

recognized entity; it is a matter of (at least partially) reconceiving the interior of the cell.

The issue that divides realists and nominalists is whether—or perhaps to what extent—such scientific work is creation or discovery. Do the scientists draw the lines, or do they discover where nature draws them? Realists maintain that nature draws the lines. The scientists and other investigators simply find what is there to be found. The problem, though, is that there are too many things there to be found.

Mereology and set theory are exceedingly tolerant disciplines. Any objects, however widely dispersed, constitute a mereological sum—an individual whole. So there is an individual made up of the world's largest frog, Fred's 2014 income tax return, Rockefeller Center, and the Brazilian navy. And except for pathologically self-referential cases, every collection of objects constitutes an extension. There is, then, an extension consisting of Napoleon's nose, the number 12, and an electromagnet in Singapore. The extension of "grue" is as determinate as the extension of "green." The entities belonging to a given mereological sum are parts of the same whole; the members of an extension are alike in being members of that extension. However gerrymandered they seem, these items are genuine individuals and extensions. So neither mere part–whole relations nor mere membership in a common extension solves our problem. Again we have an embarrassment of riches. We need a basis for, as Mary Kate McGowan (2003) says, privileging the individuals and kinds that matter.

Realists maintain that the world does the privileging (Lewis 1999). If so, there are individuals and kinds whose identity and integrity is independent of anything we know or think about them. "Green" is privileged over "grue" because "green" is a more natural kind than "grue." That is a fact about the mind-independent world. And we know that "green" is more natural than "grue" because natural science tells us so. Let's set aside the small complication that secondary qualities like "green" do not have a particularly elevated scientific pedigree and investigate the contention that the most plausible thing to say about scientific kinds is that nature marks them out.

Ordinarily, philosophers focus on kinds like water or gold, samples of which have (or at least are thought to have) a shared, stable underlying structure. That structure is supposed to explain their behavior or characteristics. But in science and in daily life, there are other kinds that seem unlikely candidates for nature's privileging.

One is *toxicity*. Many substances are toxic. Some are biological, some chemical, some physical. All are deleterious, but they affect different organisms and harm them in different ways. Some are toxic to cells, some to organs, some to organisms. Radiation, chlorine, arsenic, venom, and cigarette smoke do not have the same causes or the same specific effects. Yet, they all fall under the label "toxic." The rationale for grouping them all together and for developing the science of toxicology is our abiding interest in not poisoning or being poisoned.

Syndromes also raise difficulties. A syndrome is a constellation of signs and symptoms that collectively indicate a disorder. Often there is a single, perhaps unknown, underlying cause. Marfan's syndrome, for example, is caused by a genetic anomaly. A realist could maintain that the common cause is the reason for coalescing the signs and symptoms of Marfan's syndrome together and saying that people exhibiting those signs and symptoms have the same malady. But syndromes need have no common cause. Chronic fatigue syndrome is currently thought to result from a variety of distinct causes. The only relevant features that sufferers of chronic fatigue syndrome share are their signs and symptoms. And the only available treatment treats the undesirable symptoms. So the different causal trajectories that got the patients to their current state turn out to be medically irrelevant. Granted each case has a causal trajectory, and each particular sign or symptom has a specific cause. But the rationale for grouping them together into a single syndrome and for saying that people who exhibit them have something significant in common is that this constellation is a robust undesirable condition.

Both toxicity and syndrome are tightly linked to an interest in health and well-being. Were it not for that interest, the causal story about the behavior of each toxic agent and each chronically fatigued patient could be told. But there would be no reason to group the members of each kind together. The criterion for membership would look as arbitrary as the criterion for calling things "grue." Given our interests, however, there is nothing arbitrary about the classifications.

Still, a realist might argue, these classifications are largely medical. It is perhaps not surprising that our interest in health and well-being figures in the privileging of kinds suited to medicine. Maybe medical kinds have more in common with predicates like "fashionable" and less in common with predicates like "nucleotide" than we thought. But, he would maintain,

Constructive Nominalism

interests play no role in demarcating real scientific kinds, the kinds that figure in the hard sciences.

Again, this is not so obvious as the realist makes it seem. Biological taxonomy classifies organisms into kinds. It is plausible that members of a taxon have something biologically significant in common with one another that they do not share with members of other taxa. This is largely true at the species level. But at higher taxonomic levels, as we have seen, things get trickier. At the limit we find the biological family *Protista*, which consists of all eukaryotic organisms that are not animals or plants or fungi. It comprises, among other things, amoebas, slime molds, and bacteria, whose only common biological property (besides having cell nuclei) is that they are all *not* animals or plants or fungi. It is hard to see how nature could privilege the property of being not an animal or a plant or a fungus. The rationale for recognizing *Protista* as a family is evidently a desire for an exhaustive, simple taxonomy. To achieve that goal, taxonomists simply introduce a "none of the above" category: *Protista*.

These cases are not exceptions, they are characteristic of categorization in general. Our interests, policies, and past practices inform the construction of the categories in terms of which we represent and understand things. This is so even in the case of the realists' favorite so-called natural kinds, water and gold. When we identify water with H_2O, we ignore the fact that in any sample of water there are always some dissociated molecules.[2] We are right to ignore this fact, but in doing so, we idealize away from the way the world really is. Even the best cases cannot be construed as merely reflecting the way things are. With the growth of science and other systematic enterprises, we learn what kind of kinds suits our cognitive and practical purposes. If we want an understanding of a certain sort, we need to carve up the domain in a particular way.

As we have seen, every collection of entities, no matter how motley it seems, constitutes an extension. And each extension marks a property—that which all and only members of the extension share (see Lewis 1999). Why should we care about some extensions and properties but not others? Realists believe that nature privileges some properties (see McGowan 2003). Items that share these properties qualify as members of the same natural kind. Some scientific realists maintain that all mature sciences track natural kinds. Others are open to the idea that only fundamental physics does so. Mass, charge, and the like carve nature at the joints (see Ellis 2001). Both selective and capacious realists are apt to endorse Putnam's contention that the success of science would be a miracle if realism were not (at least approximately) true (1975:73). Constructive empiricists think otherwise (see van Fraassen 1980). At best, they believe, the success of science would be a miracle if good scientific models were not empirically equivalent to the truth. But the miracle argument is considerably weaker than its adherents believe. Lipton maintains that the argument commits the base rate fallacy (2004:196). To know whether the success of science would be a miracle if realism were false, we would have to know how unlikely it is to come up with a radically false but empirically and theoretically adequate theory. We have no way of knowing the base rate for successful science.

We need not resort to probabilities to see that the miracle argument is problematic. Sailors at sea and nomads in the desert use celestial navigation to orient themselves and find their way in trackless domains. They locate themselves by reference to the constellations, adopting a Ptolemaic perspective. Over time they refine their navigational techniques and increase their accuracy by discerning new celestial details. So celestial navigation evolves, becoming increasingly successful. Still, no one thinks that the success of celestial navigation would be a miracle if Ptolemaic astronomy were not approximately true, nor does anyone doubt that constellations human constructs that delineate patterns that appear fairly invariant when seen from Earth. For particular human projects and practices, adopting such a Ptolemaic stance is useful. That's what accounts for its success.

Might the same hold for fundamental physical properties? As Teller (2018) and Tal (2011) argue, fundamental physical magnitudes inevitably involve idealizations. "Since 1967, the second has been defined as the duration of exactly 9,192,631,770 periods of radiation corresponding to a hyperfine transition of cesium-133 in the ground state. . . . This definition pertains to an unperturbed cesium atom at a temperature of absolute zero. . . . No actual cesium atom ever satisfies this definition." Indeed, "it is physically impossible to instantiate the conditions specified in the definition" (Tal 2011:1086–1087). The second—the fundamental unit of time—is defined in terms of conditions that do not and cannot obtain.

This raises a measurement problem. Having devised and endorsed the unrealizable idealization, metrologists have to de-idealize in order to model (and eventually construct) a clock. But along what trajectories, and how far

Constructive Nominalism

253

should they de-idealize? If they go all the way back to square 1, they are left with the inchoate, varying, subjective assessments of duration that they sought to escape. For their clocks to benefit from the rigor of the definition, de-idealization should yield models that are not too far from the idealized second. That ensures that they will reflect the understanding that the underlying theory provides. And diverse realizations of the models (that is, different clocks) should yield readings that do not differ too much from one another other. That ensures that their measure is objective (Tal 2011:1091). Clocks must be built and calibrated against one another. So the satisfaction of the "not too far" requirements that Tal identifies depends on available technological means. Scientists have to de-idealize until they arrive at something that they can build and calibrate. That depends on the current level of technology. The definition of the second is acceptable only if it is susceptible of this sort of de-idealization.

A realist about duration would insist that the increasingly accurate atomic clocks are coming increasingly close to measuring duration precisely. The nominalist need not disagree. But she would insist on focusing on the role of choice in both the idealizations and the de-idealizations that go into defining and measuring the second. And she would emphasize the importance of methodological and technological factors in the choices. That is, she would insist that we not ignore the constructive element in defining the measure.

Still, it might seem that the realist could maintain that Tal's arguments just bear on measurement, not on the magnitudes being measured. There are, the realist could insist, mind-independent facts of the matter about duration, mass, or temperature. The job of the metrologist is to devise units capable of measuring those facts.

Chang (2004), however, paints a more nuanced picture, not about time but about temperature. Temperature, he shows, is a magnitude that emerged with the invention of the thermometer. Having chosen the freezing point and the boiling point of water as the fixed points that would anchor the thermometer, scientists had to decide what qualifies as freezing and boiling. Does water count as boiling when bubbles first appear, when the surface is covered with bubbles, or when the surface is bubbling rapidly? That calls for a decision. Other questions arise because the boiling point of water depends on whether it is in a glass or metal container, on the altitude at which the boiling takes place, and on how much air is in the water. These questions could be settled by stipulation. Agreement is needed, but within reasonable limits it does not much matter what the scientists agree on. The answers were not dictated by nature. Nevertheless, what we count as temperature is a product of those agreements.

The debate between scientific realists and social constructionists is often cast as a debate about direction of fit. The realist insists that acceptable scientific representations should fit the (mind-independent) facts; the social constructionist insists that the facts are whatever fit the independently accepted representations. The constructive nominalist follows Chang and suggests that fit is often an outcome of negotiation between epistemic agents, communities, and the phenomena. Temperature and the thermometer were co-constructed, each being adjusted to the demands of the other. The concept of temperature was fine-tuned to accommodate not only felt differences in warmth but also the resources available for building a reliable thermometer and the purposes to which the thermometer would be put. Thermometers had to be such that they could be calibrated to yield consistent readings in the same circumstances. Background beliefs had to be adjusted to define what makes circumstances the same. Temperature is what well-calibrated thermometers measure. Time and temperature are, I suggest, fundamental magnitudes whose contours are contrived to bring the phenomena and our scientific interests and technological capacities into accord.

Constraints on Construction

Science favors repeatable, intersubjectively agreed-upon results. So it has good reason to seek to classify its objects in terms of categories whose membership is determinate and publicly recognizable, and where an object's standing as a member of a particular kind or its instantiating a particular property is not easily destabilized. This requires that there be a lower bound on precision, that there be clarity about what differences matter, and that relatively small variations in surrounding circumstances do not affect whether a given object is an instance of a particular kind. Choice of units keys to interests as well. What sort of understanding we seek may affect whether to focus on cells, organisms, species, populations, niches, ecosystems, or something else. It might seem that these are just questions of bookkeeping. Truths about one just map onto truths about the other. But in the absence of strong reducibility, there is no reason to think that this is so. The understanding that the ecologist seeks may gloss over distinctions that are crucial in cell biology and

Constructive Nominalism

may appeal to considerations that are, from the perspective of cell biology, irrelevant.

Such considerations make constructive nominalism plausible. The interests, practices, actions, and decisions of agents figure in delineating the individuals, properties, and kinds that constitute the world.[3] This is not to say, as an idealist might, that there is nothing outside of us. But epistemic agency is involved in configuring whatever there is into individuals, properties, and kinds worth recognizing. And there are, in principle, multiple, equally acceptable but mutually irreconcilable ways of doing so.

We partition our world into toxic and nontoxic substances, colors and shades, species and genera, and a host of other groupings because in so doing we equip ourselves with the conceptual resources to pursue our various interests. Human interests overlap, compete, and range from the very rough to the extraordinarily fine-grained. So do the categories in terms of which we mark things out.

Nor are human beings the only creatures with kind-constituting interests. Many other animals carve up the world as well. Their categories may be relatively crude compared to ours, but they serve their purposes. Creatures discriminate between animals that are potential mates and those that are not, between predators and non-predators, between the edible and the inedible. Some discriminations are, no doubt, hard wired; others are the fruits of experience. Moreover, nonhuman animals sometimes draw their lines in places where we do not. Monarch butterflies are noxious to many birds. Viceroys (which look quite similar) are actually tasty and nutritious. But birds who have been sickened by eating a Monarch scrupulously avoid both, treating them as the same sort of unpalatable thing. That we think there is a difference between Monarchs and Viceroys is nothing to them. Their kinds are not our kinds.

Nevertheless, the contention that we make the categories that fix the facts seems at odds with the stubbornness of fact. The world is largely independent of our will and largely impervious to our beliefs about it. This is so, but the assumption that the stubbornness of fact is incompatible with constructive nominalism does not stand up to scrutiny. Except for what Hacking calls "interactive kinds," where the application of a descriptor to people changes the ways they experience themselves (1999:103–105), once categories are framed and criteria for their instantiation established, what if anything, instantiates those categories is independent of anything we may think or want. Consideration of obviously constructed categories shows that this is so. Human beings invented the game of baseball and set the criteria for winning. We can revise them if we like. But given the current criteria, the Lexington Lions simply do not belong to the extension of "winning team." However much we might like them to, wishing does not make it so. The stubbornness of fact then does not require naturally privileged kinds.

Still, if the entities and kinds that constitute our world are even partly a product of our efforts, then it seems, had we not made those efforts, there would be no such entities or kinds. If, for example, no one had ever dreamed up the category "dinosaur," there would have been no dinosaurs. That seems incredible. Suppose a meteor destroyed Earth after the Mesozoic Era but before rationality evolved. Dinosaurs would still have existed, even though no one ever thought of them as such. This is surely right. But it does not impugn constructive nominalism.

Although constructive nominalism is committed to the view that if there were no concept of a dinosaur, there would be no such thing as a dinosaur, it has the resources to construe the concept rigidly. Given that there is such a concept, it applies to all entities in a particular extension, under the usual range of historical, hypothetical, and counterfactual circumstances. Thus we can project the category onto the possible scenario in which rationality never evolved. In that case there would have been dinosaurs but no rational agents. We can also construe the concept nonrigidly and acknowledge that if rational agents had never contrived the concept of a dinosaur, the beasts wandering around in the Mesozoic Era would not have been members of a single kind. Had rational agents partitioned their world differently, different entities would have been members of a single kind. Maybe, for example, the carnivores would have belonged to a different kind from the herbivores, or the large ones to a different kind from the small ones, or the early ones to a different kind from later ones, or non-avian ones to a different kind from avian ones, and so forth. There are a vast number of reasonable ways of organizing the domain.

Imagine an extreme scenario: What if no rational agents ever existed and no categories contrived by rational agents applied? Wouldn't there still have been dinosaurs? Even on this scenario, there still would have been organisms that recognized conspecifics, predators, and prey. Those organisms would be able to identify potential mates with whom they could breed. They would be able to identify predators to avoid and prey to hunt. But with nothing to privilege the category "dinosaur," nothing would make it the case that at a

Constructive Nominalism

higher level of generality all the conspecifics, predators, and prey were the same kind of thing. At that level, there would be no privileged entities or kinds. All configurations of entities and all collections of them would be on a par. There would have been dinosaurs and rockosaurs and fernosaurs and cloudosaurs and every other combination. Everything, in a sense, would be there, but there would be no privileged order or organization.

There are then several ways of understanding the claims "If no one had ever contrived the concept of a dinosaur, there would have been no dinosaurs" and "If no one had ever contrived the concept of a dinosaur, there still would have been dinosaurs." Each is intelligible and, in appropriate contexts, plausible. Trouble arises only if we conflate or confuse them.

Objectivity

The question remains: If constructive nominalism is correct, how is objectivity possible? To answer this requires saying something about objectivity. The critical point to be preserved is that the world is largely independent of our beliefs and desires. Wishful thinking does not make things the way one wishes them to be, nor does believing that they are so make them so. Realism accommodates the stubbornness of fact directly. The facts are what they are, realists believe, regardless of anything we think or do. But if we consider the various ways the world is, this position looks less compelling than it first appears. The challenge is to undermine the realist intuition in a way that shows that on the nominalist alternative the world is suitably resistant to our attitudes. Let's look at some cases.

As we have seen, a game is a human contrivance. People establish rules and criteria. These rules and criteria make it possible to do such things as hit a home run. Without the institutional framework that the rules of baseball delimit, no matter what happened when a club propelled a sphere, there would be no home run. Without the rules, there would be no bats, balls, teams, or home runs. But within the framework that constitutes the game, there are such things. And given the definition of a home run, it is a determinate, objective, mind-independent fact that on September 30, 2016, Ortiz hit a home run. That he did so is not a matter of opinion, nor does wishing make it so.

It is open to us to tinker with the rules to make home runs easier or harder to get. So we can, through our rule-making activities, affect how frequently "x hit a home run" is true. This has happened in the history of baseball. But when a set of rules is in effect, whether a given player hit a home run is a matter of fact. This is so, even if the rules of a game are arbitrary. In that case, perhaps there is a sense in which the truth is arbitrary too. But still it is an objective truth in that it is independent of what we will or believe.

A game and the facts it generates may seem pretty insular. Outside of baseball, the fact that Ortiz hit a home run does not seem to matter much. The game itself seems to mark the boundaries for the significance of the fact. But not all institutional facts are so insular. Consider promising. What justifies the contention, and makes it the case, that Gareth promised Sam that he would wash the car is that Gareth's actions complied with the rules specified by the practice of promising. Only because there is such an institution does Gareth's saying, "I'll wash the car" make it a fact that he promised. Had there been no such practice, his utterance would be, at best, a prediction. In that case, he would have been under no obligation to wash the car. But if in fact Gareth promised Sam that he would wash the car, and Gareth is sufficiently trustworthy, then this fact—which the practice of promising makes possible—gives Sam a lien on the future. He is entitled to count on Gareth to wash the car, and he is entitled to a measure of confidence that Gareth will do so.

Not all practices are fact constituting. Astrology is not. There are complicated astrological rules for calculating the influence of celestial configurations on terrestrial events. Those rules might yield the verdict that today is an inauspicious day for me to take a trip. But this does not make it a fact, for despite the complicated calculations, there is no reason to believe that the correlations that astrology alleges obtain or even come close. My horoscope does not give me a lien on the future. I would be epistemically irresponsible to believe it.

Recall Rawls's (1999) distinction between justification *in* a practice and justification *of* a practice. Justification *in* a practice is determined by the rules and procedures that define and govern that practice. That Ortiz hit a home run, that Gareth promised to wash the car, that today is an inauspicious day for me to take a trip are all justified *in* the practices that generate them. Where the three differ is in the justification *of* the practices that generate them. The justification *of* the practice of baseball presumably lies in the enjoyment the practice affords to participants and fans. Insofar as people enjoy baseball, and neither the practice nor its enjoyment has significantly deleterious effects, baseball is justified. The justification *of* the practice of promising is practical. It is useful to have a mechanism whereby people can voluntarily restrict their

Constructive Nominalism

freedom in advance, so that actions can be coordinated and cooperative, temporally extended activities can be undertaken with confidence. There is, evidently, no justification *of* the practice of astrology—at least insofar as astrology makes claims about the causal structure of the world, for there is no reason to believe, and good reason to doubt, that its deliverances and methods are reliable.[4] I suggest that only practices that are themselves justified are fact constituting, for only where practices are justified is there reason to countenance their deliverances.

Consider now the case of science. The acceptability of a particular scientific finding, such as the finding that CD38 regulates oxytocin secretion (Jin et al., 2007), depends on its being justified *in* the practice of a science—that is, on its satisfying the rules, standards, and criteria that the science sets. There must be intersubjective agreement among qualified scientists that the evidence supports the finding. The methods used to generate that support must be recognized as reliable. The results must be reproducible and statistically significant. The theory that embeds the finding must display the theoretical virtues—simplicity, precision, robustness, explanatory power, and so forth. A finding that has these features will be justified *in* the science that endorses them. But to understand why these features are the features an acceptable theory ought to have, we have to consider justification *of* the practice of science.

Pretty clearly science is different from baseball and promising. As we saw, there is and need be little justification of the practice of baseball. People like playing and watching the game. That's enough. Promising is justified by its utility. The existence of the institution of promising makes certain sorts of cooperation feasible and reasonable. It might seem that justification of the practice of science is that the factors it favors are truth conducive. In that case, the findings that satisfy the standards of science are more likely to be true than those that do not. This is an appealing answer. Unfortunately, it is too simple.

Some theoretical virtues, such as robustness, have no claim to truth conduciveness. A well-established precarious finding is no less likely to be true than an equally well-established robust one. Other values, such as precision, are even more problematic. A precise finding is less likely to be true than an imprecise one. That some transmembrane glycoprotein regulates oxytocin secretion is more likely to be true than that CD38, in particular, does. So if our goal is truth, we should probably eschew precision. But to

260   Chapter 10

scrap such scientific values would be ridiculous. Rather, I believe, we should scrap the idea that the justification of the practice of favoring theories with such virtues is a matter of truth conduciveness.

I suggest that the justification of the practice of a science consists in its being as good as any available method for engendering the sort of understanding we seek.[5] Obviously, to make this non-vacuous, I need to say something about what sort of understanding this is and why it is worth having. What we want from a science is a systematic understanding of a broad range of phenomena. That understanding should be based on empirical evidence. It should explain and (depending on the science) predict the phenomena and perhaps enable us to exert control over them. This characterization might seem to say just that we want to know the way the world is and that empirical methods are, as far as we can tell, the best way of finding out. In that case, the justification of the practice of science is that it is truth conducive. This is not the whole story though. In extracting the promise, Sam wants more than that Gareth wash the car. He wants to be *entitled to confidence* that Gareth will wash the car. Analogously we want more than a cluster of empirical claims that happen to explain, predict, and perhaps enable us to control the phenomena. We want to be *confident* and to be *entitled to confidence* that science yields (or at least has the best prospect of yielding) an empirically grounded, systematic understanding of a broad range of natural phenomena. I suggest that many of the methods and theoretical virtues of science have the role of underwriting entitlement to confidence.

Comprehensive worldviews are notoriously fallible. Not only are the mythic worldviews of our ancestors riddled with errors, so are the scientific theories of our immediate predecessors. So, probably, are our own. If we want an account that we can have confidence in, we should build in resources for avoiding, detecting, hedging against, and correcting errors. Intersubjective agreement among acknowledged experts serves as a stay against carelessness, bias, and simple mistake. Reproducibility of results controls for accident. If results are reproduced and suitably varied experiments yield correlative results, the likelihood that the outcome is due to chance or to undetected interference diminishes. Statistical significance affords evidence that the result is not a fluke.

Moreover, we seek an understanding that is not easily dislodged. This requires that it be secure, stable, resilient, and robust. An understanding of a topic is *secure* to the extent that its network of commitments cannot easily

Constructive Nominalism

unravel. When there is plenty of evidence from a variety of independent sources, it is evidentially secure. Each element is tied to and both supports and is supported by a multiplicity of others. Should the evidence be sparse, it is insecure. It could easily be discredited. Parallel points hold with respect to methods, standards, and measurements. The understanding is *stable* to the extent that it can survive loss of support. If it would become unacceptable if any of the supports were discredited, it is precarious. An understanding that depends heavily on the reliability of a single test procedure is unstable. Should the procedure be discredited, the understanding disintegrates. An understanding is *resilient* to the extent that should it be discredited, readily available, acceptable revisions reinstate it. An understanding is *robust* to the extent that it is supported by multiple, independent methods (see Massimi 2022:49–73).

An idiosyncratic, irreproducible result could be true. So could a precarious, unstable, insecure, or fragile one. If *truth* is our sole objective, then science's penchant for intersubjective agreement and reproducible results seems unduly demanding. An insecure, precarious, fragile finding could be true and justified, so long as the evidence is exactly as it appears to be. Even if science seeks *justified truths*, then many of science's requirements seem too exacting. But if we seek theories that we are entitled to have confidence in, the epistemic situation is different. Then we have very good reason to endorse the values and methods of natural science. In that case, the entities, properties, and kinds recognized by science should be such that science can serve its end: to deliver an understanding of nature that we can have confidence in. What privileges particular entities, properties, and kinds is that they suit practices that are themselves justified.

Truth

What about truth? Here I draw on Putnam (1983). If a theory is consistent, it is possible to define a truth predicate for it. That yields truth-in-the-theory. If that theory satisfies the highest epistemic standards that we can devise, we have every reason to believe that truth-in-the-theory is truth. So, assuming that both our wave theory and our particle theory are ideal, we have every reason to believe that the entities they recognize are real and that the conclusions they sanction are true. But the conjunction of two truths is true. So, if (a) "Light consists of waves" is true and (b) "Light consists of particles" is true, then (c) "Light consists of waves & and light consists of particles" is true. Doesn't this land me back in the "too much of a good thing" position that I disparaged earlier? It need not. Suppose (d) "Twelve first-year students got an A" is true and (e) "Seven women got an A" is true. Then, (f) "Twelve first-year students got an A & seven women got an A" is true.

But we cannot infer that nineteen people got an A, for some of the first-year students who got an A might have been women. Because the truth of (f) might involve redundancy, not all the seemingly obvious inferences can be drawn. Similarly, I suggest, for (c). Even though, we are assuming, it is true that light consists of waves and it is true that light consists of particles, it is not true that light consists of waves and particles, for "light consists of waves" is short for "as modulated through acceptable theory A, light consists of waves" and "light consists of particles" is short for "as modulated through acceptable theory B, light consists of particles." This affords no reason to think that there exists an acceptable theory C such that as modulated through acceptable theory C, light consists of waves and particles. The laws of logic then remain universal, but in drawing inferences we need to be sensitive to redundancies. This is so, regardless of our metaphysics. It is no more a problem for constructive nominalism than it is for any other metaphysical theory.

Constructive nominalism affords a uniform treatment of truths. It countenances truths about games, fashion, ethics, and science and, at a suitable level of abstraction, gives the same account of all such truths. This, of course, qualifies as a virtue only if one thinks that there are genuine truths about games, fashion, and ethics. Otherwise, one would want to sharply distinguish such enterprises from the realm where genuine truths obtain. That, by consensus, is (or at least includes) the realm of natural science. But constructive nominalism figures in a better account of actual science than realism does, for it can vindicate the theoretical values of science. And, on a constructive nominalist account just as on a realist account, facts are stubbornly independent of our attitudes toward them. So constructive nominalism allows for objectivity. Wishful thinking is not an effective methodology. Reasoning, observing, testing, revising, and all the other hard work that goes into figuring out how things are retain their importance. What differentiates the nominalist from the realist is not her endorsement of these activities but her understanding of how much we contribute to what we find by engaging in them.[6]

11 Models as Felicitous Falsehoods

Introduction

Representations are sometimes characterized as the intensional counterparts of their referents. If so, their correctness is a function of their accuracy. The more closely a representation aligns with its referent, the better it is. But accuracy is a secondary concern. Even a highly accurate representation may be unhelpful or misleading. It might highlight irrelevant factors, represent its object at the wrong grain, adopt a skewed perspective, or swamp the interpreter with confusing and inconsequential details. In that case, it is, at best, cognitively inert; at worst, deleterious. Effective representations facilitate and direct reasoning and action. They are cognitive tools—things we think with (see Suárez 2024). That being so, they are subject to conditions of adequacy. Their epistemic acceptability depends on how, and how well, they facilitate inferences we want to draw and underwrite actions we want to perform. In designing effective representations, we trade off accuracy for adequacy. Although much of the philosophical discussion of modeling focuses on scientific models, reliance on streamlined, idealized representations is widespread. Their adequacy, in any field, depends on the extent to which they foster epistemic ends.

Models in Science

Modern science is one of humanity's greatest epistemic achievements. It constitutes a rich, variegated understanding of the natural world. Epistemology could readily accommodate its extraordinary success if that understanding were expressed in accurate representations of the phenomena. But it is not. Science couches its deliverances in models that are purposefully inaccurate—models that simplify, augment, exaggerate, and/or distort. Nor does it construe modeling as a temporary expedient. Although scientists anticipate that current models will be supplanted, they expect the replacements to be better models, not the unvarnished truth. That being so, science's pretension to epistemic eminence may look a bit dodgy. The regular, unabashed divergence from the truth of accepted, highly touted models seems intellectually suspect. How can science claim to supply understanding *of* the phenomena, if its representations are not and do not purport to be literally true of the phenomena?

Maybe my concern is misplaced. Nature is complicated; understanding it is hard. Perhaps we should just own up to our limitations and construe scientific models as approximations. Then we might contend that scientific progress consists in devising ever better approximations. Although construing models as approximations is sometimes plausible, it is not as helpful as we might hope. A variety of sorts of divergence from accuracy count as approximations. Some models are approximately accurate in all the cases that they pertain to; there's just a small margin of error. Others are accurate (or nearly accurate) in most cases but admit of a few outliers. Yet others are accurate (or nearly accurate) in most of the cases that scientists particularly care about, even though these are a relatively small subset of all the cases that the models purport to pertain to. Snell's law is a case in point. It is accurate only for isotropic media. But most media are anisotropic. Still, scientists insist, isotropic media are particularly interesting and important. Some models aggregate. Although they are approximately true of the phenomena taken collectively, they are not approximately true of any particular case. No one comes close to being an economically rational agent. But models that posit such agents are effective because individual divergences from economic rationality cancel out in the aggregate. At best then, the operative notions of approximation show a family resemblance. Hand-waving about approximation does not get us very far. If we are going to account for the epistemic success of science by saying that models are good approximations, we need to explain what makes a particular approximation and a particular sort of approximation good, recognizing that there are a variety of things we might reasonably mean by "approximation" and a variety of ways an approximation might be good.

We might think that, like the anomalies that plague current science, models are temporary expedients that will be eliminated when science reaches its goal. Maybe at the limit of inquiry, science has no need of such devices. In ultimate science, every truth-apt representation would be literally true. Paul

Models as Felicitous Falsehoods 265

Teller strongly suggests otherwise, arguing that truth and precision inevitably pull apart. A simple example illustrates his point. The velocity of sound in dry air depends on the temperature, the pressure, and the proportions of the different sorts of gas molecules that compose the air. Although we can give averages, we are in no position to precisely specify the temperature, pressure, and proportions of the different sorts of molecules in a sample of air. Nor, Teller maintains, is this just an epistemic problem. He says, "No real-world sample of air has such precise values, if only because the values would vary from place to place. So at best one is talking about the speed of sound in some idealized condition, not in the real world" (2018:278–279). Because the temperature, pressure, and proportions of different sorts of gas molecules vary slightly from one region of the sample to another, it is not true that sound travels with exact velocity v throughout the sample. To characterize the velocity of sound in air, we need to idealize, treating the magnitudes as uniform across the sample. This is what we do and what we should do. But we do not thereby mirror the actual (fluctuating) speed of sound across the air sample. Teller's point is not just that contemporary science *contains* idealized models that are strictly false. Rather, science *consists* of them.

The widespread use of inaccurate models raises important epistemological questions. First, why does science rely on models that are designed to deviate from the truth? Second, how—if at all—do such models contribute to understanding? I contend that models enable us to understand reality in ways that we would be unable to if we restricted ourselves to the unvarnished truth. The point is not just that the features that a model skirts *can* permissibly be neglected. They *ought* to be neglected. Too much information occludes patterns that figure in an understanding of the phenomena. The regularities a model reveals are real and informative. But many of them show up only under idealizing assumptions.

Angela Potochnik (2017:1–22) maintains that science's dependence on idealized models is due to the confluence of two factors: first, nature is enormously complicated; and second, human minds are limited. Both claims are true. Science is a human practice. Human finitude is not a problem for science, it is simply a condition of science.[1] Nevertheless, the emphasis on human limitations undervalues and distorts the epistemic significance of modeling. Rather than being cognitive crutches scientists regrettably lean on or heuristics that they are unfortunately fated to deploy, effective models are powerful tools that expand our epistemic range. It might seem that in saying this I concede Potochnik's point. Perhaps it is only because our range is limited that we benefit from its expansion. I'm not convinced. But before turning to models in particular, let us consider a couple of analogous cases.

Eyeglasses are called "corrective lenses" because they are devices for bringing people with defective vision to the level of visual acuity that other humans naturally exhibit. They compensate for a deficiency. Lenses in ocular microscopes and telescopes are not *corrective*. They *extend* perceptual access. With their aid, humans see what otherwise none of us could. Other instruments do more. Radio telescopes and electron microscopes, MRI scanners, and the like do not extend perception. They augment it, enabling us to detect what human perception cannot reveal. Still, Potochnik might say, such augmentation is needed only because we do not naturally have the ability to see very small things or very distant things or directly observe the internal workings of living organisms or whatever. Their value derives from our perceptual limitations. Maybe so.

Consider then a non-perceptual case. Prior to the invention of calculus, physics was hampered by the inability to compute the instantaneous rate of change of an accelerating object, the area under a curve, the exact tangent to a curve, and so on. Calculus enables us to make such calculations and many more. It massively expands our epistemic range. No one construes differential equations as mere heuristic devices, unfortunately necessary crutches, or modes of representation that will be eliminated at the end of inquiry. We fully expect ultimate science, if there is such a thing, to express its findings and make its calculations using mathematical formulas, some of which will probably be differential equations. If they are not, that will be because mathematicians have invented even more powerful tools. Newton invented calculus expressly to expand science's epistemic range—to enable it to represent phenomena in terms of continuous functions, to compute instantaneous rates of change, to make other calculations that could otherwise not be done.[2] Calculus affords a cognitively rich and rewarding way to represent and reason about things. It is not a cognitive prosthesis that enables us to accommodate our unfortunate natural failings.

I want to say the same about modeling. Like both instrumentation and mathematics, modeling expands our epistemic range. Like mathematics, it provides representations that characterize phenomena in ways that facilitate informative reasoning about them.

Models as Felicitous Falsehoods

Still, even if we recognize that instruments expand epistemic access, we might doubt that models do too. Scientific instruments are, and are designed to be, data gatherers. The more powerful our instruments, the more or better data they afford access to. But models do not gather data. Like theories, they process or interpret data. If epistemic access is fixed by our data-gathering abilities, models do not afford epistemic access. Nor, presumably, do theories. In this respect, the distinction between theories and models is idle.[3]

In any case, this objection goes too fast. First, the limits on epistemic access are not fixed by the data we can gather. What matters are the data we can process. That depends on both our data-gathering and our data-processing capabilities. If we devise new ways to process old data, we expand our access to, and often our understanding of, the phenomena they concern. This is why innovations in statistical modeling are often fruitful. They enable us to glean more information from data we already have. Models tell us what to measure, how to measure, with what precision to measure, and where to focus attention. Theories do the same. Hacking notes, "Once the theory of pulsars was in place, older astrophysical records were shown to be rich in evidence of pulsars" (1988:511). The theory transformed what was previously dismissed as noise into informative data.

Second, data are not things out there in the world that we sometimes manage to trip over. They are products of curation. Information qualifies as data only because it answers to specific requirements set by the models or theories it bears on. These requirements specify how items are to be individuated, how they are to be conceptualized, and how they are to be detected and/or measured. Seemingly relevant information is justifiably disregarded when it does not fit the demands set by the relevant models or theories. Theoretical commitments thus transform information into data.

Third, scientific instruments do more than gather data. Some process what is found in nature to generate pure samples. Others interfere with such samples in the lab. That is, they do something to the samples to produce a detectable and, it is hoped, informative change (see Hacking 1988). The access to nature that scientific instruments provide is informed and shaped by theories that dictate what is to be eliminated in purifying, and by theories of instrumentation that both enable scientists to design suitable devices and explain their affordances and limitations. Some scientific instruments are designed to register and gather data. They detect and record the information they do because they were expressly constructed to answer to theoretical requirements. This is so even for relatively simple instruments like thermometers (see Chang 2004). But it is particularly vivid in the case of complex instruments. An MRI scan is a statistical synthesis of diverse inputs that scanners register. The device was designed to register exactly the information that would contribute to the desired statistical synthesis and to omit irrelevancies. Clearly theoretical commitments lie behind the design specifications. The Large Hadron Collider uses models to synthesize outputs—to transform them into data. Completely unprocessed, its outputs would be unintelligible. Through a series of iterations, it produces tractable data.[4]

The point about instrumentation is an elaboration of the uncontroversial insight that even unmediated observation is theory laden. Observing a tree by looking at it with the naked eye requires ignoring many inputs into the visual system and synthesizing the ones that remain. It involves settling on a level of generality—seeing it as a tree rather than, for example, as an elm or a plant or an organism apparently thriving in New England or whatever. It involves taking a particular perspective on the tree—not too close, not too far, not where it is occluded by the barn. Here the counterpart of a model is a perceptual schema that equips the observer to interpret visual inputs as of a tree.

Observation, instrumentation, and modeling operate in tandem. The observations scientists can make depend on the instruments they have designed, calibrated, and validated. These in turn depend on what the relevant, available models or theories are, since they are integral to the design, calibration, and validation of instruments. Models thus inform the design of instruments that will detect the phenomena and supply data. Models process the data, yielding information that can be imputed to their targets. Because models are integral to the design of scientific instruments, they are integral to the epistemic access that empirical investigations provide. The critical point, made by Suárez (2009; 2024), Knuuttila (2011), and Hughes (1996), is that models and theories are not merely reflections of the phenomena. They are things we think with.

Effective models are powerful tools, not regrettable expedients that we fall back on only because our reach exceeds our grasp. They are powerful tools precisely because suitably designed, suitably deployed inaccurate representations are epistemic resources that foster systematic, fruitful, empirically grounded accounts of things.

Felicitous Falsehoods

Effective models are what I have called "felicitous falsehoods" (Elgin 2017:23–32). They are typically representations *of* phenomena—that is, they typically have targets. But they purposefully misrepresent those targets. Although some, being non-propositional, are not strictly false, I label them falsehoods because they misrepresent. I consider them felicitous because their inaccuracies are epistemically fruitful; they are not defects. The falsehoods are inaccurate in ways that enable them to non-accidentally provide epistemic access to obscure or occluded aspects of their targets. Not despite but because of their inaccuracy, they afford the access that they do. When inaccurate models facilitate reasoning about the phenomena, they are epistemically felicitous. In what follows, unless I say otherwise, I shall be talking about effective models. Ineffective models are ones that fail to perform or inadequately perform their epistemic functions.

Strevens (2008, 2017) takes a similar line. He maintains that effective models are epistemically legitimate because their inaccuracies are not difference makers.[5] Although real gas molecules are not perfectly spherical, their divergences from sphericity make no difference to their Boylean behavior. So when we seek to understand Boylean behavior in gases, it does no harm to treat the molecules as spheres. Similarly, although the relevant attitudes of each individual car buyer are idiosyncratic and often mutually inconsistent, collectively the divergences from economic rationality cancel out. Thus seemingly intractable idiosyncrasies are not difference makers. That being so, it does no harm to ignore them and represent each potential customer as an economically rational agent. The economic model will not, of course, provide a profile of the deliberations and actions of any individual customer. Nor will Boyle's model provide a profile of the career of any particular gas molecule. Each will, however, provide a representation of what happens in the aggregate: irrelevancies and idiosyncrasies wash out. The adequacy of a model depends on the context of inquiry where it is deployed. For a model to be epistemically effective, Strevens (2017) maintains, inaccuracies in non-difference makers are unproblematic. If this is right, the use of idealized models is epistemically permissible. Since a felicitously false model does not impute all of its seemingly relevant properties to its target, in purporting to be a model *of* that target, it need not say or implicate anything false *about* the target.

270   Chapter 11

This does not yet explain why models are epistemically advantageous. Perhaps the answer is simply that they are the best we can do. Even if we cannot devise or grasp a fully accurate, fine-grained representation of the automobile market—why Fritz bought a blue Toyota, while Sasha opted for a red Audi, and so forth—maybe we can devise a representation that will explain why overall car sales trend as they do. That would not be nothing. But, I suggest, the simplifications, distortions, and emendations of effective models do more. They afford epistemic access to patterns in the phenomena that the unvarnished truth would obscure.

Models are schematic representations that rigorously and systematically omit irrelevancies. If they did no more than omit, characterizing them as falsehoods, even in my extended sense, would be unfounded. Every representation omits something; the truth is never the whole truth. Distortions are different. They plainly falsify. Many models effect their simplifications via distortion.

Here is an example: Droplet formation is a complex hydrodynamical phenomenon, where the shape, velocity, and curvature of the fluid surface change continuously. Drawing on Batterman (2009), Potochnik describes a model of how a stream of water forms drops as it drips from a faucet. She says:

> Certain idealizations can be made in virtue of the system approaching a hydrodynamical discontinuity—that is, the breaking point—and those idealizations simplify matters considerably. . . . The water can be treated as if it were a vertical line . . . because its axial extension (length) is much greater than its radial extension (width) as the discontinuity is approached. This idealization enables a one-dimensional solution to an otherwise quite complicated set of equations (the Navier–Stokes equations). . . . Approaching the discontinuity, the relationship among surface tension, viscous forces, and inertial forces is such that acceleration due to gravity can be wholly neglected, even though it is present. . . . The axial and radial lengths both become, at the discontinuity, arbitrarily small. For this reason the shape of the fluid at the point of discontinuity is expected not to depend on the size of the system, that is, of the fluid mass and surface area. . . . These idealizations . . . demonstrate why different liquids, forming droplets in a variety of circumstances, have the same shape at the breaking point. (2017:80–81)

This constellation of simplifications is elegant, effective, and efficient. Numerous liquids share these features and thus exhibit the same behavior.

To understand the droplet formation without such a model would require solving the Navier–Stokes equations (which no one knows how to

Models as Felicitous Falsehoods

do); calculating the continuously changing shape, surface area, and velocity of the water as the drop formed; incorporating the effect of gravity, and the size, shape, and material properties of the nozzle; taking into account the actual radial and axial dimensions, the surface tension at every instant, and so forth. Potochnik rightly emphasizes that we couldn't do all this. That underscores her claim about human limitations.

My point is different. Even if we did it, that would not put us in a position to appreciate some of what the idealization discloses—that gravity, although real, is irrelevant; that a one-dimensional solution suffices; that surface tension, viscous forces, and inertial forces are equally significant. Nor would we be in a position to appreciate how the solution generalizes to other liquids, or why the generalization stops holding where it does. There is no reason to think that we could recognize that beneath the vast array of details that distinguish droplet formation in different liquids, the same pattern holds. Patterns emerge when details are neglected (see Dennett 1991). If we accommodated all the differences in detail, the patterns of droplet formation in diverse liquids would be different. Moreover, the pattern itself is, in cases like this, a matter of scientific interest (see Ambrosio 2024). That a variety of liquids exhibit the same pattern when differences in detail are ignored is important to the overall understanding of droplet formation and fluid flow. To recognize the common pattern requires ignoring specific differences. Too much information is an impediment to understanding.

Sometimes reality itself distorts. Features of the phenomena can warp underlying patterns. This is why many statistical models represent populations as infinite. Doing so swamps chance effects that afflict finite populations. To understand the roles of nonrandom factors, we devise a model where chance makes no difference. Models in population biology represent species as infinite in size. To prescind from the (currently irrelevant but nonetheless real) consequences of genetic drift, biologists construct a fictional scenario in which genetic drift would have no discernible effect. Both the fact that there are no infinite populations[6] and the fact that genetic drift is always present are sometimes irrelevant. If one wants to understand how alleles would redistribute in the absence of genetic drift, representing the population as infinite is appropriate and effective.

Models can disclose patterns that are indifferent to mechanisms or specific causal trajectories. The Lotka–Volterra model represents the interdependent fluctuations in the levels of predators and prey in a region via a pair of differential equations that describe the dynamics of interdependence. It is not peculiar to foxes and rabbits, fish in the Adriatic, or even to biological species. It cuts across disciplinary divides, holding of the relations between cheetahs and gazelles, starfish and mollusks, predatory lenders and needy borrowers, asset strippers and their targets. Assuming certain formal constraints are satisfied, whenever one population preys on another, the model applies. To be sure, there are limits. If the predators drive their prey to extinction, the pattern obviously is broken; if the predators are themselves also prey, the dynamic is more complex; and so forth. The model discloses nothing about how the pattern is realized—that is, how specific population pairs modulate their size. It discloses a regularity that holds at a level of abstraction that is indifferent to precise reproductive mechanisms. In effect, it construes species-specific reproductive mechanisms as non-difference makers. Again, prescinding from details is revelatory.

Strevens maintains that the Lotka–Volterra model is merely a schema. Because it says nothing about mechanisms—about how the model is realized in a specific predator–prey pair—it is incomplete (2008:158–159). But prescinding from mechanisms is not a defect or limitation in the model. Biologists might like to understand the mechanisms by which different species pairs modulate their populations. But the fact that the same pattern holds regardless of differences in mechanism reveals something important about the dynamics of predator–prey relations—something that would be obscured if we scrapped the model in favor of a multiplicity of independent, finer-grained accounts. An accurate, detailed representation of the modulation in populations of particular predator–prey pairs would show that reproduction rates vary in tandem. An accurate account of the mechanisms by which different pairs modulate their population size would explain how each pair does so. But because the details would vary, such an account would occlude the fact that the modulations in population size of the various predator–prey pairs display the same pattern. It would obscure the fact that differences in the ways the pairs modulate their population size make no difference.

A model that is effective for one purpose may be ineffective for another. If we seek to display a pattern that holds for a variety of predator–prey population pairs, the Lotka–Volterra model may be appropriate. If we seek to display the means by which a particular pair modulates population size, we need a model that supplies more detailed information about mechanisms. Depth of understanding is in tension with breadth of understanding. The proliferation

Models as Felicitous Falsehoods

of details that are peculiar to one realization of the model obscures the pattern that the different realizations share. To grasp a general pattern in the population dynamics of predators and prey, we need to prescind from the details.

Some models augment. Maxwell's model represents the electromagnetic field as composed of rotating vortices separated by so-called idle wheels. Via the introduction of fictional elements, the model shows how the relational structure of the production and transmission of electric and magnetic forces in an electromagnetic medium parallels the relational structure of the production and transmission of mechanical forces in a mechanical medium (Nersessian 2008:29ff.). This enables scientists to export aspects of their understanding of mechanical media to the electromagnetic realm. To read all features of the model back into the target would be a mistake. Maxwell was by no means suggesting that the electromagnetic field is littered with tiny, idle, electromagnetic wheels. They are fictional devices introduced to highlight a more abstract structure, neither mechanical nor electromagnetic, that the two systems share.

Some models exaggerate features in order to make them salient. According to Kepler's first law, Earth travels around the sun in an elliptical orbit, with the sun at one focus. Diagrammatic models of the law typically represent the major axis as considerably longer than the minor axis. If the models purported to show the actual shape of the elliptical orbit, they would be badly mistaken. In fact, the two axes are almost equal in length. But interpreted correctly, the models exemplify only the property of *being elliptical*, not the precise shape of the ellipse.[7] Under that interpretation, the models are correct. Models are symbols; they require interpretation and are vulnerable to misinterpretation.

These examples should suffice to show that effective models are felicitously false. Still, such openness to distortion, augmentation, exaggeration, and (often) extreme simplification raises the question: How do such divergences from accuracy advance, rather than inhibit, understanding?

Denotation, Demonstration, Interpretation

Models are representations designed to foster understanding by facilitating fruitful reasoning that illuminates the phenomena they concern. The liberties they take, the divergences from overall accuracy, are justified by their epistemic payoffs. A number of philosophers of science have emphasized that models are things we think with (see Suárez 2009; 2024); they are not mirrors. R. I. G. Hughes (1996) characterizes a model as a complex symbol that performs three interanimating functions: denotation, demonstration, and interpretation. His discussion is suggestive but schematic. I elaborate it to bring out features he sketches.

Denotation is the relation of the model to whatever it is a model *of*. The harmonic oscillator, being a model *of* a spring, denotes a spring; the tinkertoy model, being a model *of* a protein, denotes a protein. Demonstration consists in reasoning with the model according, as Hughes says, to "its own internal dynamic." Interpretation consists in imputing the fruits of that reasoning to the target. Denotation is familiar and straightforward. It is the relation of a name to its bearer and of a predicate to the items in its extension. Demonstration and interpretation require explication.

Demonstration: Models are subject to conditions of epistemic adequacy, which constrain and channel reasoning in a Hughesian demonstration. A model has an epistemic end: to answer a particular question or enhance understanding of a particular range of phenomena.[8] Its adequacy depends on how well it serves that end—that is, how well it facilitates informative, fruitful, nontrivial inferences about its target while impeding misleading and idle inferences. This is plainly a pragmatic matter. As Knuuttila (2011) argues, a model is a tool. Like any other tool, it is to be assessed on the basis of how well it enables users to do the job it is designed to do.

A model's internal dynamic delimits the cognitive resources it is permissible to deploy and the uses to which they can permissibly be put. Those resources include background assumptions, auxiliary hypotheses, forms of inference, taxonomies, standards of relevance and precision, and so forth. The recognition that the model is supposed to afford epistemic access to the target and answer specific questions about the target guides the choice of constraints. Descriptions, inferences, and actions that take us too far afield are sidelined.

The notion of inference operative here should be understood broadly. In addition to more rigorous forms of logical inference, a model's internal dynamic may (but need not) license analogical reasoning, associative reasoning, and/or abductive reasoning. It issues more focused licenses as well. The internal dynamic of the droplet formation model discussed above permits treating the droplet near the breaking point as a vertical line—that is, it licenses reasoning as though, near the breaking point, the droplet were two-dimensional.

Models as Felicitous Falsehoods

Moreover, reasoning according to an internal dynamic involves action as well as deliberation. Using a Newlyn–Phillips machine to represent or to figure out the effects of tweaking economic policy requires physically manipulating a flow of water, for it is by seeing how the water flows through the apparatus that one draws conclusions about the flow of money in an economy. Nor are practical inferences solely the province of material models. The internal dynamic of a purely abstract model or of a computer simulation licenses certain actions when particular results are reached. One such action is terminating demonstration—ceasing to draw further inferences. The internal dynamic tells us when to stop. We can think of a model's internal dynamic as specifying the range of permissions and prohibitions for reasoning with it.

Chains of inference are in principle endless. There is always a further conclusion that could be drawn. They proliferate in all directions. To use a model properly, we need to know what direction to take in drawing inferences and when to stop. Unrestricted inferences threaten to generate a plethora of disparate conclusions, with no obvious way to tell which ones can legitimately be imputed to the target. It follows from $pV = nRT$ that the item modeled is not a giraffe; it follows that the model is applicable in New York. Such inferences, although sound, are idle. The proper use of the model sidelines them; it takes them offline. A model has irrelevant features—features that ought to be ignored. If the demonstration phase promoted drawing valid inferences indiscriminately, these would swamp (and likely deflect) our thinking. The model must block irrelevant and unproductive inferences. How does it do so?

The answer lies in the way a model symbolizes. According to Hughes, this is a matter of exemplification. Exemplification is the mode of reference by which a symbol refers to some of its own features (Goodman 1968:52–56; Elgin 1996:170–183; Vermeulen et al. 2009). It highlights them, underscores them, makes them manifest. Exemplification is the familiar relation of a sample to whatever it is a sample of. Models are exemplars. Like commercial paint samples, they are designed to make some of their features salient. Those features may be monadic or polyadic, static or dynamic, abstract or concrete. By representing a population as infinite, for example, the Hardy–Weinberg model showcases the aspects of allele redistribution that are insensitive to random fluctuations. Exemplification is selective. To highlight some features, an exemplar marginalizes or occludes others.

The inferences licensed by a model's internal dynamic are vehicles of exemplification. They show how, for example, changes in one parameter affect changes in others, how a system evolves over time, how robust or fragile the linkages are. They disclose patterns and discrepancies that might otherwise to be hard to discern. The model does not exemplify the results of irrelevant inferences; its internal dynamic does not license them. So even when they are logically impeccable, their results are idle. By being an exemplar then, the model constrains and directs its internal dynamic toward features that can be responsibly imputed to the target.

Interpretation: Interpretation involves identifying the features exemplified at the conclusion of the model's demonstration phase and imputing them to the target. Hughesian interpretation is not literal denotation. We know perfectly well, for example, that gas molecules are not spherical. So in imputing sphericality to the molecules in the target gas—in interpreting the actual gas molecules as spherical—we do not maintain that they really are spherical. Rather we construe actual gas molecules as, in effect, spheres with distortions. In general, in imputing features of a model to a target, we construe the target as having the features exemplified by the model, distended, distorted, or overlaid by confounding features. Then we ignore the confounds as, in the circumstances, irrelevant.

A model is designed to make particular features of its target salient. Its effectiveness depends on whether the features it exemplifies illuminate the target, enabling model users to understand the phenomena it bears on.[9] By exemplifying a feature, a model affords epistemic access to it, enabling us to recognize it and appreciate its significance. The model can thereby equip us to see the target in a new light. $pV = nRT$ exemplifies the relation between temperature, pressure, and volume. It omits any mention of attractive forces. If the results of our calculations hold up to a specifiable degree of precision when we impute them to the target, we have reason to think that intermolecular forces play no significant role in the thermodynamics of the system we are investigating. We know, of course, that every material object attracts every other. So we do not conclude from the effectiveness of the model that there is no attraction. Rather, we conclude that for the sort of understanding we seek, at the level of precision that concerns us, intermolecular attraction is negligible. It is not a difference maker. To represent gas molecules as spherical then does no harm. Indeed, it helps, for by treating the molecules as spheres, we prescind from complications that would impede our understanding of the relations between pressure, temperature,

Models as Felicitous Falsehoods

and volume of a gas. This indicates that it is fruitful to think of the target in terms of the features the model exemplifies. It invites us to think of actual gases as ideal gases with distortions, of springs as harmonic oscillators with friction as a confound, of automobile buyers as rational economic agents with perhaps irrational but anyway irrelevant quirks, and so forth.

Because models omit, distend, distort, and emend, they are context and purpose relative. An inaccuracy that is illuminating in one context or for one purpose may be misleading in or for another. A psychologist interested in why teenage boys like flashy cars would not represent her subjects as rational economic agents. Such a model would elide the very features that she sought to study. To devise an appropriate model requires recognizing what factors are and what factors are not likely to be difference makers for the question one is investigating. To use a model correctly requires understanding how it functions—what phenomena it denotes, what range of features it has the capacity to exemplify, what sorts of inferences it supports and what sorts it blocks, what assumptions it makes, what scaffolding it relies on. The very same phenomenon can be modeled in mutually inconsistent ways, each of which is appropriate for a different range of problems. The nucleus of an atom can, for example, be modeled as a liquid drop or as a rigid shell. Each model exemplifies some features of the nucleus. Each facilitates some inferences and blocks others. The question for the user is which, if either, suits her current epistemic purposes.

It might seem that the best strategy would be to always use the most accurate model we can devise. Even if we are forced, for the sorts of reasons Teller adduces, to deviate from wholly accurate representations, we might think that we should deviate as little as possible. This is not as obvious as it might look. Models are epistemically effective precisely because they sideline factors that make no difference. As Knuuttila says, "There is, from a cognitive point of view, something wrong in the idea that scientific representation should aim for as accurate a representation as possible" (2011:267). To incorporate as many factors as we can and then explain that they do not matter is to invite confusion. To see this, compare gas models. Boyle's model, $pV=nRT$, is highly idealized—far too idealized for some purposes. But where pressure is low so intermolecular forces are slight, it is often appropriate. Refinements de-idealize some of its assumptions. The van der Waals model, for example, $k=n^2a/V^2$, takes intermolecular forces into account. Further refinements in the equations of state yield increasingly accurate representations. Eventually, we arrive at the virial equation:

$$pV/NkT = 1 + B/V + C/V^2 + D/V^3 + E/V^4 + \ldots$$

"which can be rendered arbitrarily precise by extending the equation indefinitely, with each added term being derivable from increasingly detailed assumptions about the intermolecular forces" (Doyle et al. 2019:356–357.).

Why then do we keep the less accurate models in our cognitive toolbox? Why didn't the virial equation supplant Boyle's, Charles's, and van der Waal's formulations, consigning them to the dustbin of the history of science? The reason is that sometimes the virial equation supplies too much information. There are cases where all we need in order to understand the gas dynamical situation is $pV = nRT$. In such contexts, the added information the virial equation provides is inert. Still, it might seem, we could in principle accommodate this by adjusting our focus. In some contexts, only a few terms of the equation exemplify. In others, more do. So we could restrict ourselves to a single model but supply different interpretations of it in different contexts.

Such a strategy might work, but it threatens to mislead. Grice's second maxim of quantity is "Do not make your contribution more informative than is required" for the purposes at hand (1989:26). When a speaker or writer conveys information, she implicates that it is relevant. In imparting the information, she makes it salient, giving her audience reason to think that in the current context it is worthy of attention. If in fact it is not worthy of attention, the implicature is unwarranted. Adducing the virial equation with its full power threatens to mislead, even if the adducer insists that we need only attend to the first few terms. Her audience is likely to wonder why the further terms were even mentioned if they aren't doing any work. Streamlined models are preferable when appropriate because they omit the irrelevancies.

Grice's maxim directly pertains to communication. I suggest, however, that the reason it is valuable for communication is that it captures something important about understanding. By eliminating or marginalizing irrelevancies, we are better able to grasp relations among relevant factors. We understand better when we respect the maxim; we do not just communicate better. The reason Grice's maxim is a conversational maxim is that it is an epistemic maxim.[10]

Streamlining is not just motivated by an aesthetic preference for desert landscapes. The omission of irrelevancies figures in a model's capacity to

Models as Felicitous Falsehoods

advance understanding of its target. The message is not "it is permissible to omit these (irrelevant) factors." It is "omitting these factors promotes the grasp of something about the phenomena that otherwise we would not, or not easily, appreciate."

Models figure in the understanding of a range of phenomena when it is epistemically fruitful to represent the phenomena as if they had the features the model imputes to them, where something is epistemically fruitful only if it either fosters or challenges the integration of the behavior of the phenomena into our evolving understanding of the world.

Every object has indefinitely many properties and stands in indefinitely many relations to other things. The vast majority of these are of no interest whatsoever. Some of the interesting and important ones are neatly labeled by our literal vocabulary. They can be directly and literally represented. Others are semantically unmarked. Prior to Newton, for example, the property shared by a falling apple, the varying heights of the tides, and the moon's staying in orbit was unmarked. If we want to recognize such properties and relations, we need to indicate them indirectly. One way to do so is by characterizing the objects that display them as-if-ishly (see Vaihinger 2009). It is as if gas molecules were spheres or as if predators were insatiable or as if the moon were falling toward the earth. Such as-if-ish representations can be epistemically important. The reason is not merely that you won't go wrong in a particular context if you think of molecules as spherical or the electromagnetic field as composed of vortices separated by idle wheels or the moon as falling. It is that the fact that you won't go wrong discloses something significant about the phenomenon. The effectiveness of the model discloses that a particular aspect of the molecule's geometry—its being somewhat spherical—is a difference maker. The model then provides emphasis and focus. It affords insight not only into what property the molecule has but which of its properties are worth registering if we want to integrate the behavior of the molecule into our understanding of gas dynamics.

I said earlier that modeling is a powerful epistemic tool. The power lies in its ability to simultaneously generate representations that afford focus and to show why that focus (even if provided as-if-ishly) is valuable. In effect, not only do models say, "This is what you should be looking at," they also say, "This is why you should be looking at it and ignoring factors that interfere with looking at it this way."

Because modeling involves Hughesian demonstration—drawing inferences and perhaps performing other actions that the representation facilitates—a science that relies on models cannot be wholly spectatorial. We cannot passively look *through* the model to see the world as we might look through a transparent pane of glass. Nor can we look *at* the model, expecting it, like a flat mirror, to reflect the phenomena without distortion. Models are epistemic tools; they enable us to grasp phenomena. Grasping is not looking. To grasp something is to be able to manipulate it to serve one's ends (Hills 2016). In scientific models, as Knuuttila (2011) and Suárez (2009, 2024) emphasize, manipulating involves inferring—reasoning with the model, taking advantage of affordances it provides. This does not preclude taking a realist stance if that means only holding that the items mature science is committed to—be they entities, structures, mechanisms, or whatever—really exist. What it does preclude is taking the view that scientific success affords direct, aperspectival access to the way the mind-independent world is anyway. Rather than enabling us to see how things are in themselves, models enable us to apprehend how things appear from a certain orientation—an orientation whose justification lies in its affording us the sort of access we need to answer the questions we want to ask and solve the problems we seek to solve, given the resources we can deploy.

The adequacy of a representation depends on its satisfying the standards of an epistemic community. Because literal vocabulary is inevitably limited, we often have to resort to nonliteral devices—metaphors, fictions, thought experiments, and idealizations. This need involve no loss of accuracy or rigor. When a nonliteral representation satisfies the standards of the relevant community, it is acceptable. In that case, it is a felicitous falsehood, for using it in inference and action promotes the community's epistemic ends.

I have argued that scientific understanding cannot, even in the long run, deliver a precise, accurate view from nowhere. Rather than being spectatorial, science is agential. Nature is complex. It displays a plethora of patterns. In their natural habitats, some are masked by irrelevancies. They emerge only under idealizations where inconsequential complications are set aside. Social structures and dynamics are also complex. We devise and deploy models and schemata to understand social arrangements and practices as well. Like scientific models, they display the pattern Hughes lays out. They denote their targets, facilitate and channel reasoning about those targets, and enable us to impute to the targets features we would not otherwise discern. Understanding

Models as Felicitous Falsehoods

involves idealized models. They are effective to the extent that the features they exemplify and impute to their targets make a difference to the characteristics and/or behavior of the phenomena we seek to understand. There is no benefit in imputing non-difference makers to targets. Indeed, such imputations can be epistemically deleterious. Too much information can impede understanding by obscuring patterns and features that streamlined models disclose. So we have reason to retain and deploy highly schematic models, even when more detailed models are available.

12 Epistemic Gatekeepers

The Problem of the Aesthetic

Theories, models, experiments, and the like are often subject to aesthetic assessment. This is uncontroversial. What is controversial is whether such assessments have any bearing on their epistemic standing. Is there an epistemically good reason to prefer an elegant experiment to an inelegant one, a beautiful theory to an ugly one, a streamlined model to one that seems more like a Rube Goldberg machine? What are we focusing on when we make such assessments?

If we seek to understand the role of aesthetic factors in science, it would be nice to have a criterion for the aesthetic. We do not have one. The history of aesthetics is littered with failed attempts to devise such a criterion. For our purposes, however, something less will do. We need only identify a few factors that are plausibly construed as aesthetic. Then we can try to determine what, if anything, they contribute. If we find a contribution, we can attempt to identify other factors that function analogously. We do not need an exhaustive list of aesthetic factors, nor do we need to determine what precisely makes those factors aesthetic.

Clive Bell maintained that the aesthetic response to works in the visual arts consists in the apprehension and appreciation of *significant form* (1913:49–71). He was concerned with significant visible forms—colors, contours, configurations, and the like. I suggest that many aesthetic responses to works in other disciplines consist in the apprehension and appreciation of disciplinarily significant forms in a logical space—scientifically significant forms in the sciences, historically significant forms in history, socially significant forms in the social sciences. The space need not be physical, and the apprehension need not be sensory. The aesthetic properties that concern us then are formal properties of such artifacts as theories, models, explanations, methods, and experiments. A form, let us say, is scientifically significant to the extent that it illuminates something that bears on the scientific acceptability of the item that displays that form. Parallel formulations hold for significant forms in other disciplines. Precisely which forms these are, of course, varies from one context of inquiry to the next. Nevertheless, the extrapolation of the idea of significant form makes it plausible that features like elegance, symmetry, simplicity, systematicity, and their antitheses are aesthetic features of scientific artifacts.

Bell's criterion is known to be inadequate. It fails to capture features of works in the visual arts that are undeniably aesthetic. No doubt, my extrapolation is equally inadequate and for the same reason. Still, I am not suggesting that significant form is a criterion for being aesthetic either in science or elsewhere. I am suggesting that significant disciplinary form is an aesthetic characteristic of some epistemic artifacts. Just as Bell's criterion enables us to identify some aesthetically important features of works in the visual arts, my extrapolation enables us to identify some aesthetically important features of works of science, history, the social sciences, and the humanities. Here I focus on science, but my points apply to other disciplinary understanding as well. My purpose is to consider what functions aesthetic factors perform. For that, I need a few plausible candidates. I do not need (and cannot provide) a full criterion. Luckily, a sketch will do.

Truth or Thereabouts

According to a familiar criterion for scientific acceptability, aesthetic factors turn out to be either irrelevant or merely instrumental. Science, it is held, has a single overriding epistemic goal. Realists maintain that the goal is truth: science is successful when it reveals the truth or something close to the truth. Instruments and methods are scientifically valuable just because and just to the extent that they are truth conducive. Norms, standards, criteria, and the like are acceptable just in case and just to the extent that they promote the discovery of truth. Constructive empiricists maintain that the goal is empirical adequacy: science is successful when it achieves empirical adequacy or something close to it. Instruments and methods are scientifically valuable just because and just to the extent that they are conducive of the development of empirically adequate accounts. Norms, standards, criteria, and the like are acceptable just in case and just to the extent that they promote the development of empirically adequate accounts. Either way, on this picture, if aesthetic factors contribute to science, their value is instrumental. They are epistemically valuable because and to the extent that they are indicative of the goal's being realized or because and to the extent that they promote its realization. To streamline discussion, I will speak as though the goal is truth; parallel arguments hold if the goal is empirical adequacy.

According to this view, the justification for preferring a beautiful, economical theory over an ugly, gerrymandered one replete with ad hoc excrescences is that the beautiful theory is more likely to be true (or close to the truth) than its aesthetically unattractive rival. An elegant experiment is preferable because it is more likely to reveal the truth (or something close to the truth) than one that proceeds incrementally by case hacking. Milena Ivanova lists Poincaré, Dirac, and Heisenberg as scientists who take beauty to be an indicator of truth. She quotes Dirac, who claims that when a theory is beautiful, "one has an overpowering belief that its foundations must be correct quite independently of its agreement with observation" (2020:88; citing Dirac 1980:40). If Dirac is right, aesthetic assessments might play a useful diagnostic role. As scientists investigate a promising theory in depth, they may come to discern hidden beauties in it. The theories whose hidden beauties they appreciate are the ones that they are increasingly confident are likely to be true or nearly so. That likelihood motivates scientists to keep working with them. And that scientists who have studied an issue in depth think that a theory is likely to be (nearly) true is reason for the rest of us to agree. We defer to their expertise. There may then be a correlation between aesthetic assessments of currently credited theories and truth-related assessments. Perhaps the fact that aesthetically sensitive scientists find a theory beautiful is a good reason to consider it prima facie acceptable. If, however, beauty and truth do not align, such a truth-centered account should hold that aesthetic judgments in science, at best, are epistemically idle; at worst, they threaten to mislead. Even if some important scientists are drawn to the beauty of their theories, it does not follow that beauty is a guide to truth (see Ivanova 2020:89).

James Ladyman groups aesthetic features like simplicity and elegance with other virtues such as coherence with background information and explanatory power. As far as I can see, the only rationale for considering them akin is that they are superempirical. That is, they are not evidence based but are nevertheless likeable features of theories. He says, "It is widely acknowledged that it is difficult to argue that simplicity or elegance are direct evidence of the truth of a theory, so realists have tended toward the view that all superempirical virtues are subsumed under explanatory power so that the solution to the underdetermination problem is IBE [inference to the best explanation]" (2012:42). Coherence with background information and explanatory power are plausibly truth conducive. Having these properties might reasonably figure in what makes an explanation the best, where the best is the most likely to be true. But even adherents of inference to the best explanation need a reason to think that the fact that an explanation has aesthetic properties like elegance and simplicity contributes to its being the most likely to be true or nearly so.

Regrettably, there is no reason to believe that a scientific representation's aesthetic features correlate with the probability of its being true. Standard aesthetic assessments do not, so far as we can tell, track truth. We appreciate harmony, symmetry, elegance, and simplicity, all of which can be found in works of art that make no pretense of being true. Many of those works are not even truth apt. Indeed, aesthetes who advocate art for art's sake would maintain that genuine aesthetic value is rightly indifferent, if not hostile, to truth. Nor is there any a priori reason to expect the phenomena to arrange themselves so as to align with our aesthetic preferences. That scientists who have studied a theory in depth think it is true is, ceteris paribus, a reason for the rest of us to think so. That they think it is beautiful is not.

Many manifestly false theories are beautiful. Aristotelian biology, for example, treats species as fixed: each species has its own essence, its own proper function, and its own distinctive telos, which both determines the distinctive good for its members and explains their behavior. This seems much lovelier than contemporary Darwinian biology, which acknowledges a large role for chance in the evolution of species, indeterminacy at the boundaries between species, a nontrivial measure of adaptive opportunism. It has nothing to say about what, if anything, constitutes the proper function of or distinctive good for the members of a species. Caloric theory, which takes heat to be a smoothly flowing fluid, seems at least as lovely as the kinetic theory of heat, which maintains that individual gas molecules career randomly about. Moreover, scientists may continue to consider an account beautiful

Epistemic Gatekeepers

even after it has been decisively rejected. Feynman, for example, initially thought that Feynman diagrams could explain why empty space is weightless. He continued to consider his account beautiful even after he refuted it (Wilczek 2016). Scientists do not automatically lose their aesthetic appreciation for the theories they leave behind.

But if aesthetic judgments in science are epistemically idle, why do we go on making them? Perhaps scientists' aesthetic assessments supervene on, derive from, or even are mere expressions of truth-related judgments. Then, in calling a theory beautiful or an experiment elegant, a scientist would be simply using an aesthetic label to characterize something that she considered scientifically estimable for other, legitimate—that is, truth-related—reasons. This seems to strip the terms from their aesthetic role. The label gets exported, but its aesthetic function is left behind. "Beautiful" seems to mean just "good of its kind" (see McAllister 1996:77–81). In that case, there is no role for genuinely aesthetic assessment in science.

Maybe aesthetic judgments in science play a heuristic role. They might function as fast and frugal ways to make preliminary assessments of theories, experiments, models, and the like. But like the heuristics that figure in psychology's System 1 thinking, they are shortcuts that not infrequently lead us astray. They may be valuable for the way they enable us (with our limited minds) to come to a conclusion. But their utility is primarily practical, and the heuristics are buggy. When you need a quick answer to "Is this likely to be true?" ask yourself, "Do you find this beautiful?" This strikes me as a dreadful idea, at least until we can find some reason to think that there is a correlation—even a loose correlation—between assessments of beauty and the likelihood of truth.

An Alternative

The problem with all of these proposals comes, I suggest, from thinking that if aesthetic factors are genuinely of value to science, it is because they somehow promote or sustain the justified conviction that the items that display them are true or nearly true. I will argue that aesthetic factors are integral to good science. They are not mere instruments, nor is their utility primarily practical. But there is no reason to think that they themselves are truth conducive. Rather, they bear on the acceptability of a scientific artifact—a theory, model, experiment, or whatever—by functioning as gatekeepers: they play a regulative role, constraining the range of considerations we are, when engaged in a particular epistemic activity, open to accepting.

The idea that factors that are not truth (or empirical adequacy) conducive are nevertheless integral to good science may seem anathema. Science, after all, seeks to understand the world. But, as I argued earlier, the understanding that science delivers is embedded in models that are not, and do not purport to be, accurate representations of the phenomena they bear on. They simplify, streamline, augment, and omit. They are not true. Some, such as Snell's law, are not even nearly true. To be epistemically acceptable, a model or theory must properly answer to the phenomena. But properly answering to the phenomena is not the same as being a literally true or accurate representation of the phenomena. Nor is it the same as being an empirically adequate one. Models that diverge from truth also diverge from empirical adequacy. The Lotka–Volterra model, for example, construes predators as insatiable. They are not. Even the hungriest shark eventually eats its fill and stops eating. Science already distances itself from the idea that acceptable models must be true or empirically adequate. So the claim that the contribution of aesthetic factors to science is not a matter of their being directly or indirectly truth (or empirical adequacy) conducive is not so alarming as it might first appear.

An understanding of a topic consists of a systematically linked body of information in reflective equilibrium. It is a network of epistemic commitments that includes not just commitments about matters of fact but also acceptance of models, idealizations, and thought experiments that are known to diverge from truth, as well as of methods, norms, and standards that are not truth apt. Literal truths are not privileged. To be justified in believing that the Higgs boson exists (or, for that matter, that the Loch Ness monster exists) requires a commitment to the methods, norms, standards of evidence and rigor, as well as to the background assumptions that figure in establishing its existence. Since understanding is, in the first instance, understanding of a topic or range of phenomena—not understanding of an individual matter of fact—the various elements of a network must be mutually supportive. Ivanova too takes aesthetic factors to be integral to scientific understanding. She maintains that the explanation is grounded in our cognitive makeup (2020:98–100). Human beings cognitively favor accounts that display aesthetically positive features. I'm not convinced. Works of art that are ugly, disordered, complicated, unbalanced—Francis Bacon's portraits, Goya's

Epistemic Gatekeepers

Disasters of War, Bosch's *Garden of Earthly Delights*, Pollock's *Convergence*—all contribute to understanding. Rather than thinking that aesthetic factors like beauty, elegance, simplicity, and order are factors favored by human psychology, I suggest that they are factors favored by scientific understanding—the sort of understanding that scientific communities seek and design their practices to achieve. When an aesthetic factor such as symmetry or elegance is integral to a network of scientific commitments in reflective equilibrium and could not be eliminated from the system without threatening, weakening, or undermining the system's reflective equilibrium, then if that system is, by current standards, at least as good as any available alternative, the factor is epistemically justified. We need not think that the explanation is that the feature is favored by cognition as such.

Even if beauty is not exclusively in the eye of the beholder, it is hard to characterize. Nor is it wholly a matter of significant form, however broadly construed. Lopes maintains that beauty is a determinable, with elegance, symmetry, and so forth as its determinants (2018:128). I will focus on several aesthetic determinants that figure in science—ones that seem to be largely, if not entirely, matters of form: symmetry, systematicity, simplicity, and elegance. I am not saying that these factors are aesthetic per se. The only one with any claim to that status is elegance. But followers of Lopes might well take them all to function as determinants of beauty in science. Whether or not this is so, their function in scientific acceptance is aesthetic. A model displaying symmetry, an experiment's being simple, a consideration's weaving seamlessly into a network of mutually supportive commitments are the sorts of things that make them epistemically attractive in science.

Symmetry

Symmetry is a matter of structural invariance. A symmetrical item—an object or a law—retains its structure under transformations. Thus, for example, a cubical block displays rotational symmetry in that it retains its shape when rotated. The members of a collection that share a symmetry have the same structure and retain that structure under the same transformations. They constitute an equivalence class. With regard to the transformation in question, they are interchangeable (van Fraassen 1989:243). Things are not symmetrical tout court. They are symmetrical in one respect or another. So in devising a theory or model, questions arise: What sorts of symmetries are we interested in? Under what sorts of transformations should the objects of interest retain their structure? These questions depend on the sort of theory or account we want to devise. The answers mark out our categories and shape our theorizing. The symmetries that figure in science are not like rocks; we do not just stumble over them as we go about our lives. They are products of decisions. As I argued earlier, we configure the domain, deciding what structures we want to preserve and under what transformations we want to preserve them. Maybe, for example, we seek to preserve rotational symmetry but have no interest in color invariance. Then the block would count as symmetrical even if its faces were different colors.

It is no accident that scientific models display specific symmetries. We design them to do so. In designing a model or partitioning a domain in such a way that certain symmetries are displayed, we tentatively commit ourselves to the view that those symmetries are scientifically significant forms. What is the basis of their significance? We have no reason to think that a symmetrical model is more likely to fit its target than an asymmetrical one or that it is likely to fit its target better than one that lacks the symmetry in question. A critical question is how symmetries and asymmetries affect epistemic decisions.

Buridan's ass found himself equidistant from two identical, equally nourishing and tasty bales of hay. With respect to choice worthiness, the two bales were symmetrical. Nothing favored one over the other. Symmetry paralyzed the poor beast. Rather than starve, let us hope, he would eventually choose arbitrarily—perhaps favoring the one on the right, even though there was absolutely no reason to prefer it to the one on the left. If he did, we could not understand his decision. We could understand that he chose—indeed, understand why he had to choose. But we could not understand why he chose the bale on the right. There was no reason. This is unsatisfactory, particularly if we are in the business of studying ass psychology. We might give him a bye if this were a one-off choice. That much arbitrariness in the mental life of an ass we might be willing to tolerate. But if he found himself in the same situation on multiple occasions and chose the bale on the right significantly more frequently than the one on the left, we would be apt to insist that there must be a reason. There must be something about the hay on the right or the ass's psychology, we would be apt to think, that accounts for the difference. There *must* be a tiebreaker. Then we would embark on a quest for hidden variables.

Another case: A fair coin is rotationally symmetrical. In the flip of such a coin, nothing favors heads over tails. Suppose we observed a single coin

Epistemic Gatekeepers

flipped seventeen times in a row. Each time, the coin came up heads. We know that in an infinite sequence of tosses of a fair coin, there is bound to be an interval where the coin comes up heads seventeen times in a row. Nevertheless, we would almost surely suppress that knowledge and judge that the coin was biased. We expect a fair coin to come up heads about half the time, even in a short run of tosses. Should it not, we again seek a hidden variable—something that biases the coin.

There is, of course, no guarantee that we will find the hidden variable we seek. Maybe there is none to be found. My point is that in cases like these, we treat the perception of symmetry and the perception of asymmetry differently. If the bale of hay on the right were fresher or nearer or composed of different sorts of hay from the one on the left, we would probably consider that a symmetry-breaking difference and think we understood the ass's behavior. If the coin came up HTTH THTH TTHH THTT H (which specific sequence is, after all, no more or less probable than a sequence of seventeen heads in a row), we would judge the coin fair. But the fact that the all-heads sequence makes the coin *look* asymmetrical arouses our suspicions. We do not think we understand how it happened. Something, we think, needs to be explained.

The conviction that a range of phenomena display a certain sort of symmetry is an initially tenable commitment (see Elgin 1996:101–106). In constructing a system in reflective equilibrium, that conviction has a slight and defeasible claim on our epistemic allegiance. We need a reason to give it up. Such reasons are often easily found. We may discover that preserving a commitment proves too costly. Maybe the only way to preserve the conviction that the coin was biased or that bales on the right are more desirable than ones on the left is to invoke occult forces for which we can find no independent evidence. Being unwilling to do that, we conclude that the coin was in fact fair and that the ass's several choices were mutually independent and arbitrary—that our suspicions about the cases were unwarranted. Our quest for hidden variables in cases like these is evidence of our commitment to symmetry. We may find that we have to rethink the situation and our conviction that symmetry was broken where we thought it was. Or we may have to weaken our commitment to symmetry and recognize a measure of pure chance in nature. But we abandon the commitment reluctantly. We prefer symmetry-preserving accounts. Symmetry, I suggest, is an aesthetically pleasing feature. Our preference for it affects our behavior in accepting or rejecting certain findings.

292   Chapter 12

Systematicity

As we have seen, White distinguishes between a chronicle and a history (1965:222–225). A chronicle is just a list of unconnected, established facts about an historical episode. It makes no mention of dependence relations between facts, nor does it attempt to impose any order on them. A history is organized. It links the various facts to one another, displays dependence relations, imposes an explanatory framework in terms of which it makes sense that things played out as they did. Pretty clearly, the epistemic value of a history of a given event is correlated with the number and perceived importance of the facts in the corresponding chronicle that it accounts for. The history need not mention every fact listed in that chronicle. But, one way or another, the history should accommodate the important facts on the list. A history of the siege of Stalingrad that left it a mystery why so many people died would be unsatisfactory.

Something similar holds in science. A science seeks systematic, integrated, mutually supportive accounts of the phenomena it investigates. It eschews danglers. A series of independently established truths about a domain would qualify as a natural chronicle. But a scientific account would be unsatisfactory if it remained mysterious how those truths hang together.

Unaccommodated, isolated, but seemingly significant truths are considered problematic. Toward the end of the nineteenth century, celestial mechanics entertained a variety of increasingly strained hypotheses to accommodate the apparently anomalous precession in the perihelion of Mercury. But even at their most desperate, scientists did not advocate accepting:

All planets except Mercury have regular Newtonian orbits; Mercury is just different.

Although the contention is true, was confirmed by the observational evidence, and is not a Gettier case, it is unacceptable without an explanation of exactly why Mercury is different.[1]

Unexplained danglers are unacceptable. A question is considered open until putative exceptions to a proposed answer have been explained or explained away. Unless apparent exceptions can be shown to be irrelevant or misconceived, they are in one way or another expected to be woven into the scientific account. To be sure, there are anomalies—seemingly relevant issues that currently defy explanation or incorporation into our best available accounts. But anomalies are construed as outstanding debts. A scientific account is unsatisfactory to the extent that it lacks the resources to pay its debts. Moreover, once a phenomenon has been woven into an acceptable account, scientists are typically satisfied. Nothing more needs to be explained. Why Mercury's orbit is irregular was an outstanding problem for Newtonian mechanics. Why Mercury's orbit is regular is not a question for general relativity. Indeed, the question hardly makes sense. After all, given the theory, what would you expect?

Enthusiasm for systematicity runs deep. We want our fabric of scientific commitments to be tightly woven. Physicists seek a Grand Unified Theory (GUT) out of a dissatisfaction with having to admit multiple fundamental forces into their ontology. It would be aesthetically more pleasing if there were only one. We disparage case hacking as a way of establishing a hypothesis because the fact that the hypothesis can be shown, one by one, to apply to each individual case does not to our satisfaction show why it applies to all of them. The significant form of an acceptable scientific account consists in its being an interwoven (preferably tightly interwoven) fabric of epistemically interdependent commitments that has few, if any, danglers. Ideally, the GUT would be "of unlimited validity and intuitively satisfying in its completeness and consistency" (Weinberg 1992:6)

Simplicity

Simplicity is complicated. There are numerous dimensions along which a theory or model might be simple or complex. Ockham's razor is traditionally formulated as a principle of ontological simplicity: Do not multiply entities beyond necessity. But, as Sober (2015) points out, there are a multiplicity of razors, each seemly worthy of adoption. There is no obvious general reason to favor ontological simplicity over, for example, axiomatic simplicity, syntactic simplicity, or inferential simplicity. Ontological simplicity consists in a commitment to a minimal number of primitive entities or a minimal number of primitive kinds of entities. Axiomatic simplicity consists in containing a minimal number of axioms, fundamental laws, or postulates. Syntactic simplicity is a matter of the number and intricacy of the basic syntactical units that figure in laws or axioms. Inferential simplicity is a function of the length of the derivations from the fundamental laws to other commitments of the theory. No doubt this list could be extended and the distinctions sharpened. For our purposes, the point is that these are all reasonable, objective, attractive features of theories and models. Moreover, they seem to have little to do with truth or empirical adequacy. There is, on the face of it, no reason to think that a theory with a minimal number of primitives or one that admits of streamlined inferences is more likely to be true than one that has a greater number of primitives or one that requires long, convoluted inferences. Nor is there any reason to think that the prospects of empirical adequacy are enhanced by either sort of simplicity.

It might seem then that our preference for simplicity is grounded in intelligibility or tractability (see De Regt 2017:246–251). Simpler theories and models are easier to handle. This is a pragmatic asset. If so, at least one reason why a simple theory is to be preferred over a complex one is that we are less likely to make mistakes in thinking with it or acting on it. The preference then is an accommodation to our human cognitive limitations. This would align with Potochnik's (2017) position. But sometimes simplicity comes at the cost of intelligibility. Classical propositional logic needs only one primitive: the Sheffer stroke. In terms of ontological simplicity, Sheffer stroke logic wins hands down. But it is hard to do logic using only the Sheffer stroke. The proofs are long, many steps seem counterintuitive, mistakes are apt to be made. As a practical matter, it is preferable to formulate propositional logic using at least two, and typically three, primitives. Standard formulations of propositional logic are, moreover, inferentially simpler than Scheffer stroke logic. Proofs are shorter, more direct, and more intuitive. If intelligibility and tractability are our guides, a somewhat complicated theory may be more attractive than the simplest theory we can devise.

In the quest for simplicity, we face trade-offs. We readily sacrifice a measure of one sort of simplicity in order to gain a different sort. Thales hypothesized that everything is water (see Aristotle §983b27–33). Della Rocca (2020) thinks that reality is one and indivisible. Albert (1996) suggests that reality consists in a single physical object—the universal wave function or, alternatively, the single universal particle. All three theories achieve an admirable measure of ontological simplicity. Albert and Della Rocca contend that, au fond, there is only one thing; Thales contends that there is only one sort of thing. All face the enormous burden of accounting for the ways the one seems to configure itself into armadillos and artichokes, volcanos and viruses, molecules and mountains, and so forth. The laws they need to invoke are apt to be exceedingly complex to account for the manifest diversity of appearances at every level—if there is in reality only one thing or one sort of thing.

Epistemic Gatekeepers

Still, we are not willing to jettison simplicity altogether. We reject theoretical excrescences. An electric current is standardly described as a flow of electrons. What would be wrong with elaborating this claim, saying that the electrons are propelled by tiny, undetectable gremlins nudging the electrons along with hockey sticks? The answer is obvious: the standard account is simpler. The introduction of gremlins with hockey sticks is completely superfluous. It contributes nothing to the tenability of the account that contains it. Because it contributes nothing, the evidence in favor of the original thesis does nothing to support the gremlin addendum, nor do we have any independent evidence for it. That being so, we not only have no reason to accept the gremlin hypothesis; we have excellent reason to reject it.

This is an easy case, for the hypothesis in question is idle. It adds nothing. Things get trickier when an added hypothesis adds something but not enough. If, to explain the behavior of Buridan's ass, we had to introduce an additional, otherwise inert and undetectable psycho-magnetic ψ force that operated only when an organism was equidistant between two equally appetizing alternatives, we would resist and decide on balance that we should rest satisfied with the view that the ass's choices were arbitrary. In the absence of independent evidence of the existence of such a force, we would deem the cost in added complexity too high for the payoff it promises.

Ptolemaic astronomy was committed to the view that celestial objects display a complicated pattern of circular motions involving epicycles, equants, and deferents. It was crucial that all the motions be circular. Later astronomers, including Galileo, thought the pattern was unduly complicated. The price Ptolemaic theory had to pay to preserve the geocentric framework was too high. Still, the theory was as simple as it could be if it was to accommodate the phenomena within a geocentric system. What the Ptolemaic astronomers never did, and would have been utterly unjustified in doing, was add additional (perhaps undetectable) motions to increase the number of dimensions along which celestial objects moved in circles. Science favors a minimalist aesthetics. Rococo additions have no place.

In both the case of electron flow and the case of Ptolemaic astronomy, it might seem that concerns for simplicity are presumptively truth conducive. The complexifying hypotheses we reject are ones that, on the evidence available, we have no reason to consider true. But sometimes we favor simplicity over truth. We devise models that prescind from truth in order to achieve simplicity. Boyle's law—$pV = nRT$—falsifies the behavior of gas molecules by ignoring the attractive forces between them. For certain purposes, this is reasonable. The attractive forces are too weak to make a significant difference to the phenomena we seek to accommodate, hence too insignificant to figure in the explanations we seek to provide or the predictions we seek to make. Moreover, a more realistic model, such as the virial equation, is apt to occlude patterns that an idealized model exemplifies. As we have seen, where attractive forces between gas molecules are negligible, it is not just reasonable but advisable to neglect them— to omit them from the representations in terms of which we understand the phenomena. Streamlined models often embody an understanding of the phenomena precisely because they are simplified. They omit what is negligible—what are not difference makers—thereby enabling us to apprehend and appreciate the significance of what is non-negligible—the difference makers.

Science's preference for simplicity is rather inchoate, with different sorts of simplicity trading off against one another and different types of simplicity predominating in different contexts (see Sober 2015). Still, we reject rococo accounts. Why? During his scientific realist phase, Hilary Putnam ventured the hypothesis that the laws of nature could be no more complicated than differential equations (1975:309). It might seem that a hard-nosed scientific realist, as Putnam was at the time, should endorse such a claim only if God assured him that the challenge of figuring out the way the world is would not be too difficult for humans to meet, much as a teacher might assure her students that the test will not be too difficult if they study. We have no such assurance. So if we have to back Putnam's hypothesis with truth-conducive reasons, it is unwarranted. We should admit that we have no clue how complicated the laws of nature are likely to be. Alternatively, however, we might take Putnam's proposal to be an aesthetic constraint on acceptance. Perhaps scientific investigators have such a strong aversion to mathematical complexity that they balk when the equations get too complicated. Deeming complicated formulas unacceptably ugly, they consider the investigation that led to them still to be open and seek a way to either replace them by or reduce them to something simpler. Again, there is no guarantee that they will succeed. But such an aesthetic preference would explain their efforts to come up with simpler laws.

Elegance

Elegance in science is a fusion of effectiveness and economy. Effectiveness is an instrumental matter. There is something we seek to achieve; when we are effective, our efforts pay off. The Miller–Urey experiment was effective in that it sought to demonstrate and succeeded in demonstrating that amino acids emerge from reactions of chemicals—ammonia, methane, hydrogen, and water—which were believed to be plentiful on Earth in prebiotic times. The four chemicals in plausible proportions were sealed in a chamber that was subjected to occasional sparks, which mimicked the effect of lightning. Over several days, a series of chemical reactions occurred, eventually yielding thirteen amino acids (see Ball 2005:124–138). The elegance of the experiment lies in its simplicity, which engenders a sense of inevitability. Given the experimental design, it appears, nothing but the four chemicals could account for the production of the amino acids. The result was reached with no extraneous theoretical, computational, or material factors. The background assumptions were clear and uncontroversial. The design was straightforward. Indeed, the experiment is so simple that it almost looks like a nerdy high school science project. The question remains: How does the *elegance* of the experiment figure in its epistemic function? Why aren't elegance and effectiveness simply two distinct properties that the experiment instantiates? The answer, I suggest, is this: the experiment is convincing largely because it is elegant. Whether the experiment yields an *accurate result* is independent of its elegance; whether it yields one that *ought to be reflectively endorsed* is not.

There is no obvious reason why an elegant experiment is more likely to yield a truth, or an important truth, than an inelegant one. Nevertheless, elegance is an epistemically advantageous property. An elegant experiment makes manifest what it achieves and how it achieves what it does. It exemplifies its scientific contribution. An inelegant experiment might disclose the same truth, but we would have a harder time recognizing that or appreciating how it did so. Science seeks more than a suitably accurate representation of the facts. For a result to be scientifically creditable, there must be reason to accept it. Having generated amino acids in the way it did, the Miller–Urey experiment affords strong evidence that amino acids are products of reactions among the chemicals enclosed in the apparatus. There seems to be no other way they could have come about. This affords evidence, albeit not conclusive evidence, that amino acids emerged in nature in a particular way *if* the prebiotic atmosphere consisted of those particular chemicals in roughly the proportions reflected in the experiment. It also affords somewhat weaker evidence that amino acids emerged in that way *if* the atmosphere contained additional components, but not ones that would significantly interfere with the sequence of chemical reactions the experiment displayed.

Science wants more than just evidence for its hypotheses. It also wants *grounds for confidence* that the evidence is creditable. It wants reason to think that the findings it endorses are suitably accurate reflections of the facts. These grounds are, in effect, higher-order evidence that the first-order evidence is good. The more extraneous the factors, the more reasonable it is to wonder whether we should be confident that the result bears on the hypothesis in the way that we think it does. By, as far as possible, eliminating the extraneous—that is, by ensuring that only factors that are relevant to the outcome are actually integral to the experiment—we strengthen our reason to believe that the experiment reveals what we take it to reveal. If an inelegant experiment is sufficiently convoluted, it invites the worry that unappreciated confounding factors, rather than the hypothesis being tested, account for the result. An elegant experiment is more illuminating. Either the result is readily integrable into a currently accepted network, or it manifestly poses a challenge to the acceptability of that network. Moreover, the experiment's elegance underwrites confidence not just in its result but also in the experimental design. Because the design is so streamlined, chemists can tell exactly what is going on. They don't have to worry about hidden variables because there seems to be nowhere for variables to hide.

An elegant scientific artifact—an experiment, argument, model, or theory—*shows* why it works. It highlights the basis of its effectiveness. So, I suggest, a major epistemic contribution of scientific elegance lies in its advancing science's *understanding of itself*. An epistemically elegant artifact makes the basis of its claim to trustworthiness manifest.

Optional Stops

So far, my discussion has been largely descriptive. Scientists seek symmetry, they favor simplicity, they strive for systematicity, they appreciate elegance. I've urged that symmetry, simplicity, systematicity, and elegance are aesthetic properties, but I haven't yet said much about how they contribute to scientific understanding. One contribution has already been hinted at.

Epistemic Gatekeepers

Deviations from the ideals of symmetry, simplicity, and systematicity often demand explanation. They pose a problem that ought to be addressed. Conformity to the ideals needs no explanation. Indeed, any attempt to explain cases that conform to our desiderata is apt to look a bit weird. No one asks why things behave as expected, even when aesthetic factors figure in the expectations. There is, then, an imbalance in our demands for explanation.

Scientists evidently observe Nozick's Optional Stop Rule (1981:2). Inquiry has no foreordained stopping point. If we reach a result that we find sufficiently implausible or unpalatable, it is always open to us to conclude that there must be something wrong with the investigation, the reasoning process, or the background assumptions that led to it. This, I suggested, is what scientists who follow Putnam's recommendation do when they arrive at a mathematically complex scientific law. A scientist's reason for exercising the Optional Stop Rule may itself be aesthetic. Weinberg (1992) suggests that the quest for a GUT is at least partly motivated by a dissatisfaction with the idea that there is a multiplicity of mutually irreducible fundamental physical laws. Rather than accept a result and move on, it is open to us to decide to investigate a matter further. This may lead us to refine our methods, question our presuppositions, or look for hidden variables.

There is, of course, no guarantee that further investigation will lead to a conclusion we like better. It may be that the original inquiry was impeccable and the result, although unpalatable, was correct. Convinced that each individual event has a cause, a physicist might invoke the Optional Stop Rule and insist that there must be a reason why a particular radioactive particle was emitted at time t and another seemingly identical one was not. Rather than recognize that radioactive decay is stochastic, he might insist that scientists should keep looking. Once a no-hidden-variable theorem has been proven, it might seem, we reach a natural stopping point. What more could we want? The difficulty is that the Optional Stop Rule still applies. Rather than accept the conclusion that there is no hidden variable, we can pursue the suspicion that there is something wrong with the theorem's proof. Such resistance is not illegitimate, even when there is nothing more to find.

The point is not that we are always right in our assessments. It is rather that a plausible, palatable result is, in large part because of its plausibility and palatability, apt to be deemed acceptable. Then inquiry with respect to that question ends. A sufficiently implausible or unpalatable result is apt to spark further inquiry. How could it be? The result is treated as a challenge. More work needs to be done.

The Optional Stop Rule might itself seem untenable. It seems rather unfair, even prejudicial, to subject objectionable results to greater scrutiny than attractive ones. But actually, the rule and the treatment it prescribes are reasonable. The results that strike us as plausible and palatable are ones that are readily integrated into accounts in or approaching reflective equilibrium. And they strengthen the accounts they are integrated into.

As we have seen, a constellation of epistemic commitments is in reflective equilibrium when its components are reasonable in light of one another, and the constellation as a whole is at least as reasonable as any available alternative in light of our antecedent commitments. In constructing a system of thought, we begin with whatever antecedent commitments (beliefs, norms, methods, goals, etc.) we take to bear on the topic we seek to understand and the sort of understanding we seek to achieve. Our starting points are a motley crew. They are apt to be mutually inconsistent and, even where consistent, non-cotenable. They are likely to be gappy, failing to cover matters we think they should cover. They frequently contain errors, omissions, incongruities, and other cognitive infelicities. They are not acceptable as they stand. But they encapsulate our current understanding of the phenomena, the ways to investigate them, the norms that bear on acceptability, and so on. So we start with them. The starting points are initially tenable commitments. We correct, excise, extend, and amend them to bring them into accord. Although none of our initially tenable commitments is completely unrevisable, commitments have different degrees of epistemic inertia. We are more reluctant to abandon some than others. Even this reluctance to abandon an entrenched commitment is defeasible. Should it turn out that the price of retaining a commitment is too great—that is, should it turn out that we can only retain it by revising or rejecting other commitments that we think are collectively more worthy of acceptance—we will abandon even a commitment with considerable inertia (see Elgin 1996:101–111).

Aesthetic considerations may be initially tenable. At the outset we might, like Quine, simply have a fondness for desert landscapes. No matter. If we think that theories should be simple or that laws should disclose symmetries, we can begin by favoring accounts with those features. In that case, we build into our theorizing a bias against complexity and asymmetry. This may seem question begging, but it is not. Or anyway, it is no more question begging than a bias in favor of comprehensiveness or evidential adequacy. In any case, the bias is readily overrideable should it prove too costly. Alternatively, at the outset, we may lack aesthetic biases, having no preference for elegance, simplicity, and the rest. Then our early attempts at adjudication will be indifferent to such aesthetic considerations. But should we find that simplicity, elegance, and the like are characteristics of the accounts we ultimately endorse and discover that the closest competitors that lack those characteristics are, on the whole, less tenable, these aesthetic characteristics acquire the status of initially tenable commitments that (at least weakly) constrain future theorizing. Considerations that display these characteristics will, ceteris paribus, have an easier time gaining admission than those that lack them.

Still, to say that we would like our theory to display a certain aesthetic profile does not assure that we will get what we want. Every component of the system is subject to review. We may find that, for example, the cost of simplicity is too high. If we have to sacrifice, say, a considerable measure of comprehensiveness or evidential adequacy to satisfy our current criterion for simplicity, we have a prima facie incentive to revise the criterion. We may abandon it completely, or we may restrict its scope, concluding that, for example, the payoff for recognizing a multiplicity of distinct elements has advantages that ontological monism cannot match. On the other hand, scientific developments may strengthen and reinforce an aesthetic commitment. Arguably, symmetry was a rather peripheral epistemic value in classical physics. In quantum mechanics, it has moved to center stage.[2]

Ivanova argues that the aesthetic preferences displayed in science are grounded in our cognitive makeup. Human understanding incorporates them (2020:99). I disagree. Plenty of excellent works of art traffic in complexity, asymmetry, disorder. I suspect that our cognitive makeup is more tolerant and flexible than her discussion suggests. Nevertheless, the discussion has shown that we are within our epistemic rights to initially prefer scientific theories, models, and experiments that display particular aesthetic profiles. When we do so, aesthetic factors play a regulative role. They make no claim to track truth or empirical adequacy. Rather, their function as gatekeepers is to shape our accounts, encapsulating our evolving ways of framing our understanding so that considerations that we accept are in reflective equilibrium and therefore worthy of our reflective endorsement.

Coda

Gatekeepers set parameters, sketching contours on the appropriateness of taxonomies, measurements, standards of rigor, methods of investigation, and modes of inference. They are grounded in a discipline's conception of the sort of understanding it seeks to supply. They constrain the commitments a discipline or inquiry is prepared to countenance. Although I have focused on aesthetic gatekeepers, not all gatekeepers are aesthetic.

Some are practical. A research protocol might mandate that the results of a particular investigation be registrable in terms of magnitudes that available devices can measure. Such a constraint would limit the precision of the inquiry's findings, since exceedingly small differences cannot be measured. The requirement of measurability might hold across the discipline, being grounded in the discipline's conception of itself as engaged in experimental inquiry. The restriction to currently available instrumentation might be justified by the need, in this particular case, for timely results. The refusal to entertain options that would require the development of novel measuring devices would be a matter of expediency. Both would bear on the adequacy of findings, not their accuracy.

Other gatekeepers are prudential. A discipline might prefer to err on the side of caution, given that errors are in the offing. Thus criminal law endorses a version of Blackstone's principle, which holds that it is better for ten guilty parties to go free than for one innocent person to be convicted. The ten-to-one formula is negotiable, but the standing requirement of proof beyond reasonable doubt embeds a justifiable bias against wrongful conviction. Pharmacological research takes a similar tack in its protocols for testing on human subjects. It first requires demonstration that a new drug does no harm. Only then does it go on to test whether it does any good. Figuring out where the lines should be drawn—what gates should be guarded—is a higher-order obligation of epistemic communities. And different communities, with their different challenges and goals, quite reasonably draw the lines in different places.

Gatekeepers influence choices about the taxonomies that communities of inquiry devise and deploy to demarcate the items in their domains. They constrain the models, idealizations, and other (hopefully) felicitous falsehoods that the community contrives, as well as the modes of inference it countenances to foster the sort of understanding it seeks. They figure in the framing of specifications for the devices that a community of inquiry will rely on.

Gatekeepers do not operate in isolation, nor are they impervious to criticism, revision, or rejection. Advancement of understanding is a product of dynamic, epistemically iterative processes. Not only can various independently reasonable expectations and aspirations conflict with one another,

Epistemic Gatekeepers

but the world pushes back. Inferences we favor may prove inapplicable or ineffective; methods we rely on may be inadequate, given the standards we set for ourselves. The phenomena we seek to understand may resist being described in a particular way for particular purposes at a favored level of precision. Then the community has to reevaluate its ends and means, resources and strategies, background assumptions about the phenomena, the methods for investigating them, the goals of the investigation, and the standards by which it evaluates the inquiry and its results. This can happen on a global or a local scale. On a global scale, it often seems drastic; physics needs to rethink its commitment to the universality of the causal law when confronted with the phenomenon of stochastic radioactive decay. On a local scale, it is mundane, requiring troubleshooting and tweaking when glitches emerge or when inquiries do not pan out. Typically, there are no algorithms for how to revise. But global or local, reevaluation is creditable only if conducted by epistemically autonomous agents who are willing and able to formulate and advocate for their views.

In our quest for understanding, we do not merely want the account we arrive at to be correct; we want to be confident that it is correct, and we want a right to that confidence. Correctness is a matter of answering to the facts. How, how closely, and in what respects an account answers to the facts is a function of the sort of understanding the account purports to supply. Epistemically responsible choices are integral to deciding this. Confidence is grounded in epistemic agents having reason to think that an account appropriately answers to the facts. A right to confidence is grounded in that reason standing up to scrutiny—that is, in its satisfying the standards set and reflectively endorsed by legislating members of a realm of epistemic ends. We have, I think, no right to think that our theories are true or empirically adequate. But when they are, and we have reason to think that they are, in reflective equilibrium, we have reason to think that they are worthy of endorsement. Reflective equilibrium is a realizable epistemically estimable standard of correctness. It does not claim or promise permanent acceptability, but taken together, commitments in reflective equilibrium configure an ecological niche within which inquiry can thrive. Still, because conditions change as new techniques and methods emerge, new questions arise, and new problems appear, the growth of understanding is dynamic. A viable understanding provides a platform for further inquiry.

Notes

Chapter 1

1. On this interpretation, models without targets—e.g., biological models of species with four sexes—might be construed as fictions.

2. I am grateful to Jan-Willem Romeijn for pressing me on this point.

3. I am grateful to Glen Anderau for raising the issue of epistemic sustainability.

4. It is not clear whether this is an objection to Hills's position or merely an elaboration of it. She mentions only the drawing of deductive and probabilistic conclusions, but nowhere does she say that only these sorts of inference satisfy her requirements.

5. There is a hedge. Hills says "in the right circumstances." So, she could argue that the objections I raise simply show that the circumstances are not right. This is implausible. The inarticulate mechanic and the prelinguistic child are not hampered by circumstances. It's not just that they are in the wrong place. Rather, they are incapable of explaining because of something inherent about their condition. The ballet master's inability is different. His problem is that the language lacks the resources for communicating the extremely fine-grained distinction between stances that he needs to convey.

Chapter 2

1. I thank Jonathan Matheson for helping me clarify this point.

2. I am grateful to Jonathan Adler for this point.

3. This is a variant on Pritchard's example where the house burns down due to a short circuit. I do not know enough about short-circuit fires to make my case using his example.

4. I disagree with Grimm about this, but I will not dispute the matter here.

5. There is a caveat. It may be that certain strands in one's understanding should be, and should remain, subliminal. Perhaps, for example, networks or portions of

networks consisting of what Gendler (2010) calls "aliefs" are valuable precisely because they operate below the threshold of cognizance. Maybe they can function as effective heuristics because they skirt the added burdens that come with cognizance. The integration of aliefs into a network that contains beliefs may be valuable. It enables agents to deploy fast and frugal heuristics, which work well enough often enough. The mere fact that they are embedded falsehoods is not an objection. Propositional elements in a network are acceptable if they are true enough (see Elgin 2017). The issue that concerns us here is not the fact that aliefs are apt to be false but rather that they are apt to be subliminal. Subliminal commitments exact an epistemic cost. They may convey benefits that compensate for their cost. But they are risky. To the extent that the understanding of a topic is subliminal, it is not susceptible to the sort of intentional correction or improvement that conscious understanding is.

6. I thank Dorothy Edgington for encouraging me to clarify this point.

7. The realm of ends is an ideal. The legislators are idealized agents. There is no suggestion that it is realized in any actual legislature. I thank Tom Kelly for insisting on this.

8. I thank Jonathan Matheson for helping me articulate this point.

9. The student's relation to his exam is not an ideal model here. If the exam asks the student to discuss, say, the importance of the Battle of Trafalgar to the Napoleonic Wars, he has a strong incentive to say something. Even if he is fully aware that his answer would not satisfy the relevant epistemic standards, he also recognizes that saying something is better than saying nothing. If he is not entirely clueless, then giving an answer that strikes him as plausible or one that is unacceptably vague may at least get him partial credit. I set this point aside.

10. I am grateful to Jonathan Matheson and Kirk Lougheed for constructive comments on an earlier version of this chapter.

Chapter 3

1. Rather than Grice's supermaxim of quality "Try to make your contribution one that is true," I think the proper maxim is "Try to make your contribution true enough" (see Elgin 2017:25–26). Here, either Grice's maxim or my revision will do.

2. A dispositional construal handles the Robinson Crusoe worry. Being along on a desert island, Crusoe has no opportunity to communicate. If an agent had to act virtuously in order to be virtuous and truthfulness is a virtue, it would seem to follow that Crusoe is not truthful. Nor, of course, is he untruthful. On my view, whether he is truthful depends on how he would behave if he were to communicate. We may never know. But his moral/epistemic character determined by what he would do, not by what we have evidence of his doing.

Notes

3. Hawley takes it for granted that serious, unhedged statements of fact are assertions, and that truth is the norm of assertion. I have argued otherwise (Elgin 2017:18). So in my view someone could count as trustworthy even if neither she nor anyone else was in a position to vouch for the truth of what is said. Still, if we bracket the commitment to the view that seriously saying that p (where one's purposes are cognitive) is always bound by the truth norm and bracket the idea that a trustworthy representation need be propositional, Hawley's conception of trustworthiness meshes with mine.

4. "Presumptively," since the recipient might have other commitments which calls the information into doubt.

Chapter 4

1. As I construe the traditional criterion, reliabilism, knowledge-first epistemology, and virtue epistemology all qualify as traditional. For reliabilists, the reliability of the method by which the agent connects with the facts constitutes the justification, for that is what aligns the belief with its truth maker. For virtue theorists, the virtue of the epistemic agent secures the alignment, thereby supplying the justification. For knowledge-first theorists, the mere connection between the mental state and the fact it pertains to suffices. Such a construal obviously elides differences that for other purposes are important.

2. This evades Kripke's dogmatism paradox, since it both provides a good basis for serious inference and action and disincentivizes dismissing putative counterevidence (2011:43).

3. Parallel points can be made for fields such as logic and mathematics, where the support for a contention is not empirical. In such cases, we speak of support by reasons rather than by evidence. For my purpose, here the difference is negligible. To keep matters simple, I focus on the empirical case.

4. Here, I use "cold," "warm," "mild," and "cool" to characterize the relevant phenomenally identified properties. I resist using the terms "heat" and "hot," since our current concept of heat is too tightly tied to our concept of temperature, which only emerged toward the end of the inquiry. It is widely recognized that variations in phenomenally identified properties of warmth and coolness are distinct from variations in temperature.

5. It was nowhere near as simple as this sketch suggests. Even figuring out when water qualifies as freezing or boiling was not obvious. Then, there is the problem of altitude—boiling in Mexico City takes place at a different point than boiling in New Orleans. I leave all these complications aside, as they do not bear on my problem. See Chang (2004) for an excellent discussion of the complexities.

6. Their claim to objectivity consists in their being impartial and consistent (see Elgin 2017).

Notes

7. I mention astrology from time to time, not because I think it poses problems for my account but rather because critics regularly raise it as a challenge to my views. It is not.

8. Kripke does not endorse the dogmatic stance but rather merely sketches the rationale for it.

Chapter 5

1. Outsiders can also hold communities and their members accountable. So, for example, those affected by smoking can hold the tobacco industry accountable for its lies about the dangers of smoking. Here, I neglect accountability to those outside a community.

2. My norms of accommodation closely mirror what DiPaolo (2019) calls norms of compensation. I prefer "accommodation" because "compensation" suggests that they function to make up for a shortfall. That is not obvious. I believe that non-omniscience and fallibility are often assets rather than liabilities.

3. My discussion of this tragedy draws extensively on Barrotta and Montuschi (2018). They emphasize that it was both an ethical and an epistemological disaster. Although I focus on the epistemological dimension, the two are inextricably linked.

4. In response to political pressure, studies continued after the dam was complete. Geologists (or perhaps politicians) became convinced enough of the danger to construct a bypass to protect the town closest to the dam. But tragically, they apparently never entertained the possibility of a massive landslide.

5. This is a requirement on how they relate to one another when functioning as legislating members of a realm of epistemic ends. It is not a requirement on their political situation in the broader community. So, in principle, a viable epistemic community could be embedded in a repressive regime. I thank Tom Kelly for pushing me to make this clear.

Chapter 6

1. It does not follow that the acquisition of testimonial knowledge is never costly. It may take considerable effort to interpret and/or assimilate the information imparted (see Malfatti 2020). My point here is simply that it is not always costly.

2. There is a slight complication. A person can authorize an agent to speak and act on her behalf. The agent can make promises on behalf of his client. The client is at fault if those promises are not kept. In effect, the agent adopts the persona of the client in making the promise. Since he is authorized to do so, the client is bound by the promise.

Notes

3. Jennifer Lackey (2008) presents a vignette that is supposed to be a counterexample to the claim that the testifier must accept (or in her words believe) that p. In chapter 8, I provide a different interpretation of her case. But even if she is right that testimony does not require that the testifier believe (or accept) the information she conveys, it is hard to see why anyone would or should accept testimony if she didn't think the testifier believed (or accepted) it.

Chapter 7

1. Zagzebski speaks of belief where I would speak of acceptance. In discussing her position, I use her terminology. It is clear that she thinks that beliefs underwrite inference and action when our ends are cognitive. So, the difference in terminology should not be misleading.

2. I am grateful to Samuel Elgin for this point.

3. A rebutting defeater is a consideration that discredits the evidence for p or the adequacy of that evidence. An undercutting defeater discredits p itself.

Chapter 8

1. These equivalences are obvious idealizations. No two people are exactly alike in these respects. I follow the literature in making the idealizations, since my point is that differences of opinion are possible, even when there is considerable overlap in abilities and backgrounds.

2. I model this discussion on the controversy over the Divje Babe flute. I focus on whether it is a Neanderthal artifact, although there is also considerable controversy about whether it is a flute.

3. Beebee (2018) shows how this plays out in philosophical disagreements.

4. See Samuel Elgin (2015) and James Lenman (2000) for arguments that show that consequentialists are never in a position to know or reasonably believe that a given action is good. They know that it is good if and only if it maximizes utility. Because causal chains are endless, we cannot know which action satisfies that requirement.

5. This example was suggested to me by Samuel Elgin.

Chapter 9

1. Whether informants are default entitled to believe testimony or need evidence of the trustworthiness of the testifier is a matter of controversy (see Burge 1993; Fricker 1987). I set this debate aside, since both sides agree in taking it that felicitous testimony transmits entitlement.

310 Notes

2. Evidentially adequate theories, as I use the phrase, are empirically adequate, given the available evidence. This is narrower than full empirical adequacy. But it suffices for their teaching to be epistemically responsible.

3. Scholars can understand theories that they do not reflectively endorse. So, Stella's stance vis-à-vis the theory of evolution might be like a classicist's stance toward ancient Greek religion. Each takes a body of commitments as the object to be understood. Each understands what the theory is committed to. But neither understands the phenomena the theories purport to concern via the theories in question (see Malfatti 2019).

Chapter 10

1. Hacking (1995) focuses on classifications that display looping effects. The availability of such a classification affects the behavior of those so described by influencing their self-conception and behavior. This introduces complications for the social sciences, since the regularities in the behavior of the people they study before and after a particular label becomes available are apt to diverge. Although I focus on modes of classification that do not display such looping effects, it is worth noting that the cases Hacking discusses—multiple personality, hysteria, autism, etc.—create challenges for a realist who would consider such categories natural kinds.

2. I am grateful to Paul Teller for this example.

3. This suggests that the position I call "constructive nominalism" may be close to or even the same as Dupré's "promiscuous realism" (see Dupré 1993). If there is a difference between our views, it lies in the importance I assign to Chang's point that even fundamental magnitudes are products of negotiation.

4. Maybe astrology as an amusement has the same sort of justification as baseball does. It's fun, and it's not hurting anyone. In that case, it plays no epistemic role and makes no cognitively serious claims about causal influences.

5. "Available method" is significant. It allows that scientific methods evolve over time. When, for example, statistics develops more refined methods, science demands better statistics before countenancing a result.

6. I am grateful to Paul Teller for discussions of a previous draft of this chapter and for access to his unpublished papers, as well as to an anonymous referee for *Synthese*, who gave me valuable advice on how to strengthen the argument.

Chapter 11

1. I am grateful to an anonymous referee for encouraging me to emphasize the extent to which Potochnik and I agree.

Notes   311

2. Leibniz invented calculus to solve metaphysical problems. That, too, was an expansion of our epistemic range.

3. I am grateful to Otávio Bueno for articulating this worry.

4. I am grateful to Anjan Chakravartty for help articulating this point.

5. Strevens maintains that all difference-makers in the natural sciences are causal (2008:158–159). Although models may "black box" details about actual causes, they are committed to it being the case that whatever is in the black box makes a difference to the causal explanation of the phenomena in question. I disagree. I believe that there are noncausal patterns, regularities, and relations of dependence, independence, and interdependence in nature. Science looks for them too (see Lange 2017).

6. Here, I will register an objection to construing models as approximations. A large finite number is not approximately infinite. It is as far from infinity as a small finite number is. So, construing these models as approximations strictly makes no sense.

7. I am grateful to Douglas Marshall for this example.

8. It may have other ends—perhaps practical ones—as well. Here, I am concerned only with its epistemic ends.

9. The illumination may be indirect. The model may exemplify something that exemplifies something else that . . . bears on the target phenomena. Or the model may have no target. A biological model of how a species with five sexes would redistribute its alleles is one. Such a model is effective if the features it exemplifies shed indirect light on actual allele distribution—for example, if the features it exemplifies makes salient something subtle about how alleles redistribute or fail to redistribute in a species with fewer sexes. We may discover something important about the actual case by considering a suitably constructed counterfactual case.

10. I am grateful to Tamer Nawer for encouraging me to clarify this point (see Nawer 2021).

Chapter 12

1. As it turned out, of course, Mercury is not relevantly different. The Newtonian theory that construed it as different was wrong.

2. I am grateful to Steven French for this point.

References

Adler, Jonathan (1981). "Skepticism and Universalizability." *Journal of Philosophy* 78 (3):143–156.

Adler, Jonathan (2002). *Belief's Own Ethics*. Cambridge, MA: MIT Press.

Albert, David (1996). "Elementary Quantum Metaphysics." In J. T. Cushing, Arthur Fine, and Sheldon Goldstein (Eds.), *Bohmian Mechanics and Quantum Theory: An Appraisal* (pp. 277–284). Dordrecht: Kluwer.

Ambrosio, Chiara (2024). "Diagrammatic Thinking, Diagrammatic Representation, and the Moral Economy of Nineteenth Century Science." In Cornelis de Waal (Ed.), *Oxford Handbook of C. S. Peirce* (pp. 261-27). Oxford: Oxford University Press.

Aristotle (1941). *The Basic Works of Aristotle*. Edited by Richard McKeon. New York: Random House.

Audi, Robert (2006). "Testimony, Credulity, and Veracity." In Ernest Sosa and Jennifer Lackey (Eds.), *The Epistemology of Testimony* (pp. 25–49). Oxford: Oxford University Press.

Ball, Philip (2005). *Elegant Solutions*. Cambridge: Royal Society of Chemistry.

Barnett, Zach (2019). "Philosophy Without Belief." *Mind* 128 (509):109–138.

Barrotta, Pierluigi, and Eleanora Monduschi (2018). "The Dam Project: Who Are the Experts?" In Pierluigi Barrotta and Giovanni Scarafile (Eds.), *Science and Democracy* (pp. 17–33). Amsterdam: John Benjamin Publishing Company.

Batterman, Robert (2009). "Idealization and Modeling." *Synthese* 169 (3):427–446.

Beebee, Helen (2018). "Philosophical Skepticism and the Aims of Philosophy." *Proceedings of the Aristotelian Society* 118 (1):1–24.

Bell, Clive (1913). *Art*. New York: Frederick A. Stokes.

Berker, Selim (2013). "The Rejection of Epistemic Consequentialism." *Philosophical Issues* 23:363–387.

Bourget, David, and David J. Chalmers (2014). "What do Philosophers Believe?" *Philosophical Studies* 170 (3):465–500.

Brun, Georg (2020). "Conceptual Re-Engineering: From Explication to Reflective Equilibrium." *Synthese* 197 (3):925–954.

Burge, Tyler (1993). "Content Preservation." *Philosophical Review* 102 (4):457–488.

Callan, Eamonn, and Dylan Arena (2009). "Indoctrination." In Harvey Siegel (Ed.), *Oxford Handbook of Philosophy of Education* (pp. 104–121). Oxford: Oxford University Press.

Chang, Hasok (2004). *Inventing Temperature: Measurement and Scientific Progress.* Oxford: Oxford University Press.

Chang, Hasok (2022). *Realism for Realistic People.* Cambridge: Cambridge University Press.

Christensen, David (2007). "Epistemology of Disagreement: The Good News." *Philosophical Review* 116 (2):187–217.

Code, Lorraine (1991). *What Can She Know? Feminist Theory and the Construction of Knowledge.* Ithaca, NY: Cornell University Press.

Cohen, L. Jonathan (1992). *An Essay on Belief and Acceptance.* Oxford: Oxford University Press.

Davidson, Donald (1980). "Actions, Reasons, and Causes." In *Essays on Actions and Events* (pp. 3–20). Oxford: Oxford University Press.

Della Rocca, Michael (2020). *The Parmenidian Ascent.* Oxford: Oxford University Press.

Dellsén, Finnur (2021). "Rational Understanding: Toward a Probabilistic Epistemology of Acceptability." *Synthese* 198 (3):2474–2498.

Dennett, Daniel (1991). "Real Patterns." *Journal of Philosophy* 88 (1):27–51.

De Regt, Henk (2017). *Understanding Scientific Understanding.* New York: Oxford University Press.

De Regt, Henk, and Dennis Dieks (2005). "A Contextual Approach to Scientific Understanding." *Synthese* 144 (1):137–170.

Descartes, René (1979). *Meditations on First Philosophy.* Indianapolis: Hackett Publishing Company.

Dewey, John (1916). *Democracy and Education.* New York: Free Press.

DiPaolo, Joshua (2019). "Second Best Epistemology: Fallibility and Normativity." *Philosophical Studies* 176 (8):2043–2066.

References

Dirac, Paul (1980). "The Excellence of Einstein's Theory of Gravitation." In M. Goldsmith, A. Mackay, and J. Woudhuysen (Eds.), *Einstein: The First Hundred Years* (pp. 41–46). Oxford: Pergamon Press.

Dormandy, Katherine (2018). "Epistemic Authority: Preemption or Proper Basing?" *Erkenntnis* 83 (4):773–791.

Doyle, Yannick, Spencer Egan, Noah Graham, and Kareem Khalifa (2019). "Non-Factive Understanding: A Statement and a Defense." *Journal for General Philosophy of Science* 50:345–365.

Dupré, John (1993). *The Disorder of Things: Metaphysical Foundations of the Disunity of Science*. Cambridge, MA: Harvard University Press.

Elga, Adam (2007). "Reflection and Disagreement." *Nôus* 41(3):478–502.

Elgin, Catherine (1988). *With Reference to Reference*. Indianapolis: Hackett Publishing Company.

Elgin, Catherine (1996). *Considered Judgment*. Princeton, NJ: Princeton University Press.

Elgin, Catherine (2010). "Persistent Disagreement." In Richard Feldman and Ted Warfield (Eds.), *Disagreement* (pp. 53–67). Oxford: Oxford University Press.

Elgin, Catherine (2017). *True Enough*. Cambridge, MA: MIT Press.

Elgin, Catherine (2020). "The Mark of a Good Informant." *Acta Analytica* 35 (3):319–331.

Elgin, Samuel (2015). "The Unreliability of Foreseeable Consequences: A Return to the Epistemic Objection." *Ethical Theory and Moral Practice* 18 (4):459–466.

Ellis, B. D. (2001). *Scientific Essentialism*. Cambridge: Cambridge University Press.

Feldman, Richard (2014). "Evidence of Evidence is Evidence." In Jonathan Matheson and Rico Vitz (Eds.), *The Ethics of Belief* (pp. 284–300). Oxford: Oxford University Press.

Firth, Roderick (1981). "Epistemic Merit: Intrinsic and Instrumental." *Proceedings and Addresses of the American Philosophical Association* 55 (1):5–23.

Foley, Richard (2001). *Intellectual Trust in Oneself and Others*. Cambridge: Cambridge University Press.

Frankfurt, Harry (1971). "Freedom of the Will and the Concept of a Person." *Journal of Philosophy* 68 (1):5–20.

Fricker, Elizabeth (1987). "The Epistemology of Testimony." *Proceedings of the Aristotelian Society Supplement* 61 (1):57–83.

Fricker, Miranda (2009). *Epistemic Injustice*. Oxford: Oxford University Press.

Friere, Paolo (2000). *The Pedagogy of the Oppressed*. London: Bloomsbury Academic Press.

Fumerton, Richard (2010). "You Can't Trust a Philosopher." In Richard Feldman and Ted A. Warfield (Eds.), *Disagreement* (pp. 90–110). Oxford: Oxford University Press.

Galilei, Galileo (2015). *Sidereus Nuncius*. Translated by Albert Van Helden. Chicago: University of Chicago Press.

Gardiner, Georgi (2022). "Attunement: On the Cognitive Virtues of Attention." In Mark Alfano, Colin Klein, and Jereon de Ridder (Eds.), *Social Virtue Epistemology* (pp. 48–72). London: Routledge.

Gendler, Tamar (2010). "Alief and Belief." In *Intuition, Imagination, and Philosophical Methodology* (pp. 255–281). Oxford: Oxford University Press.

Goldberg, Sanford (2010). *Relying on Others*. Oxford: Oxford University Press.

Goldberg, Sanford (2013a). "Disagreement, Defeat, and Assertion." In David Christensen and Jennifer Lackey (Eds.), *The Epistemology of Disagreement: New Essays* (pp. 167–189). Oxford: Oxford University Press.

Goldberg, Sanford (2013b). "Defending Philosophy in the Face of Systematic Disagreement." In Diego Machuca (Ed.), *Disagreement and Skepticism* (pp. 277–294). New York: Routledge.

Goldberg, Sanford (2018). *To the Best of Our Knowledge*. Oxford: Oxford University Press.

Goldman, Alvin (1986). *Epistemology and Cognition*. Cambridge, MA: Harvard University Press.

Goldman, Alvin (1999). *Knowledge in a Social World*. Oxford: Clarendon Press.

Goldman, Alvin (2018). "Expertise." *Topoi* 37 (1):3–10.

Goodman, Nelson (1968). *Languages of Art*. Indianapolis: Hackett Publishing Company.

Goodman, Nelson (1983). *Fact, Fiction, and Forecast*. Cambridge, MA: Harvard University Press.

Goodman, Nelson, and Catherine Elgin (1988). *Reconceptions in Philosophy and Other Arts and Sciences*. Indianapolis: Hackett Publishing Company.

Grasswick, Heidi (2018). "Epistemic Autonomy in a Social World of Knowing." In Heather Battaly (Ed.), *The Routledge Handbook of Virtue Epistemology* (pp. 196–201). London: Routledge.

References

Greco, John (2020). *The Transmission of Knowledge*. Cambridge: Cambridge University Press.

Grice, Paul (1989). "Logic and Conversation." In *Studies in the Way of Words* (pp. 1–144). Cambridge, MA: Harvard University Press.

Grimm, Stephen (2006). "Is Understanding a Species of Knowledge?" *British Journal for the Philosophy of Science* 57 (3):515–535.

Grimm, Stephen (2012). "The Value of Understanding." *Philosophy Compass* 7 (2):113–117.

Grimm, Stephen (2014). "Understanding as Knowledge of Causes." In Abrol Fairweather (Ed.), *Virtue Epistemology Naturalized: Bridges between Virtue Epistemology and Philosophy of Science* (Synthese Library, vol. 366, pp. 329–345). Cham: Springer.

Grimm, Stephen. (2016). "Understanding and Transparency." In Stephen Grimm, Christoph Baumberger, and Sabine Ammon (Eds.), *Explaining Understanding: New Essays in Epistemology and Philosophy of Science* (pp. 212–229). New York: Routledge.

Hacking, Ian (1983). *Representing and Intervening*. Cambridge: Cambridge University Press.

Hacking, Ian (1988). "On The Stability of the Laboratory Sciences." *Journal of Philosophy* 85 (10):507–514.

Hacking, Ian (1995). "The Looping Effect of Human Kinds." In Dan Sperber, David Premack, and Anne James Premack (Eds.), *Causal Cognition: A Multidisciplinary Debate* (pp. 351–394). Oxford: Clarendon Press.

Hacking, Ian (1999). *The Social Construction of What?* Cambridge, MA: Harvard University Press.

Hannon, M. (2019). *What's the Point of Knowledge?* Oxford: Oxford University Press.

Harris, Paul (2012). *Trusting What You Are Told: How Children Learn from Others*. Cambridge, MA: Harvard University Press.

Hawley, Katherine (2019). *How to be Trustworthy*. Oxford: Oxford University Press.

Henderson, David, Terence Horgan, Matjaz Potric, and Hanna Tierney (2017). "Nonconciliation in Peer Disagreement: Its Phenomenology and Rationality." *Grazer Philosophische Studien* 94 (1–2):194–225.

Heydrich, Wolfgang (1993). "A Reconception of Meaning." *Synthese* 95 (1):77–94.

Hills, Allison (2016). "Understanding Why." *Noûs* 50 (4):661–688.

Huemer, Michael (2011). "Epistemological Egotism and Agent-Centered Norms." In Trent Dougherty (Ed.), *Evidentialism and its Discontents* (pp. 17–33). Oxford: Oxford University Press.

Hughes, R. I. G. (1996). "Models and Representations." *PSA* 2:S325–336.

Hunter, Michael (1982). *The Royal Society and Its Fellows : 1660–1700.* British Society for the History of Science Monographs, Vol. 4. Oxford: Alden Press.

Ivanova, Milena (2020). "Beauty, Truth, and Understanding." In Milena Ivanova and Steven French (Eds.), *The Aesthetics of Science* (pp. 86–103). New York: Routledge.

Jäger, Christoph (2016). "Epistemic Authority, Preemptive Reasons, and Understanding." *Episteme* 13 (2):167–185.

Jäger, Christoph, and Federica Malfatti (2020). "The Social Fabric of Understanding: Equilibrium, Authority, and Epistemic Empathy." *Synthese* 199:1185–1205.

James, William (1948). "The Will to Believe." In *Essays in Pragmatism* (pp. 88–109). New York: Hafner Publishing Company.

Jin, Duo, Hong-Xiang Liu, Hirokazu Hirai, Takashi Torashima, Taku Nagai, Olga Lopatina, Natalia A. Shnayder, Kiyofumi Yamada, Mami Noda, Toshihiro Seike, Kyota Fujita, Shin Takasawa, Shigeru Yokoyama, Keita Koizumi, Yoshitake Shiraishi, Shigenori Tanaka, Minako Hashii, Toru Yoshihara, Kazuhiro Higashida, . . . Haruhiro Higashida (2007). "CD38 Is Critical for Social Behaviour by Regulating Oxytocin Secretion." *Nature* 466 (7131):41–45.

Johnson, Casey (2018). "Just Say 'No': Obligations to Voice Disagreement." *Royal Institute of Philosophy Supplement* 84:117–138.

Kant, Immanuel (1981). *Grounding of the Metaphysics of Morals.* Indianapolis: Hackett Publishing Company.

Kelly, Thomas (2005). "Epistemological Puzzles about Disagreement." In Tamar Szabo Gendler and John Hawthorne (Eds.), *Oxford Studies in Epistemology* (vol. 1, pp. 167–196). Oxford: Oxford University Press.

Khalifa, Kareem (2017). *Understanding, Explanation, and Scientific Knowledge.* Cambridge: Cambridge University Press.

Kitcher, Philip (1990). "The Division of Cognitive Labor." *Journal of Philosophy* 87 (1):5–22.

Klein, Morris (1972). *Mathematics Through the Ages.* Oxford: Oxford University Press.

Knuuttila, Tarja (2011). "Modelling and Representing: An Artefactual Approach to Model-Based Representation." *Studies in History and Philosophy of Science* 42 (2):262–271.

Kripke, S. (2011). "On Two Paradoxes of Knowledge." In *Philosophical Troubles* (pp. 27–51). Oxford: Oxford University Press.

Kuhn, T. (1970). *The Structure of Scientific Revolutions.* Chicago: University of Chicago Press.

References 319

Kvanvig, Jonathan (2003). *The Value of Knowledge and the Pursuit of Understanding.* Cambridge: Cambridge University Press.

Lackey, Jennifer (2008). *Learning from Words.* Oxford: Oxford University Press.

Ladyman, James (2012). "Science, Metaphysics, and Method." *Philosophical Studies* 160 (1):31–51.

Lange, Mark (2017). *Because without Cause.* Oxford: Oxford University Press.

Leibniz, Gottfried (1989). "Principles of Nature and Grace." In Roger Ariew and Daniel Garber (Eds.), *Philosophical Essays* (pp. 206–213). Indianapolis: Hackett Publishing Company.

Lenman, James (2000). "Consequentialism and Cluelessness." *Philosophy and Public Affairs* 29 (4):342–370.

Lewis, David (1983). "Truth in Fiction: Postscript." In *Philosophical Papers* (vol. 1, pp. 276–278). Oxford: Oxford University Press.

Lewis, David (1999). "New Work for a Theory of Universals." In *Papers in Metaphysics and Epistemology* (pp. 8–55). Cambridge: Cambridge University Press.

Lewis, David (2000). "Academic Appointments: Why Ignore the Advantage of Being Right?" In *Papers in Ethics and Social Philosophy* (pp. 187–200). Cambridge: Cambridge University Press.

Lipton, Peter (2004). *Inference to the Best Explanation.* London: Routledge.

Lipton, Peter (2009). "Understanding Without Explanation." In Henk deRegt, Sabine Leonelli, and Kai Eigner (Eds.), *Scientific Understanding: Philosophical Perspectives* (pp. 43–63). Pittsburgh, PA: University of Pittsburgh Press.

Longino, Helen (1990). *Science as Social Knowledge: Values and Objectivity in Scientific Inquiry.* Princeton, NJ: Princeton University Press.

Longino, Helen (2013). *Studying Human Behavior.* Chicago: University of Chicago Press.

Lopes, Dominic McIver (2018). *Being for Beauty.* Oxford: Oxford University Press.

Malfatti, Federica (2019). "Can Testimony Generate Understanding?" *Social Epistemology* 33 (6):477–490.

Malfatti, Federica (2020). "Can Testimony Transmit Understanding?" *Theoria* 86 (1): 54–72.

Massimi, Michaela (2022). *Perspectival Realism.* Oxford: Oxford University Press.

McAllister, James (1996). *Beauty and Revolution in Science.* Ithaca, NY: Cornell University Press.

McGowan, Mary Kate (2003). "Realism, Reference and Grue." *American Philosophical Quarterly* 40 (1):47–57.

Mill, John Stuart (1978). *On Liberty*. Indianapolis: Hackett Publishing Company.

Morgan, Mary, and Margaret Morrison (Eds.) (1999). *Models as Mediators*. Cambridge: Cambridge University Press.

Nagel, Thomas (1974). "What Is It Like to Be a Bat?" *Philosophical Review* 83 (4):435–450.

Nawer, Tamer (2021). "Veritism Refuted? Understanding, Idealization, and the Facts." *Synthese* 198 (5):4295–4313.

Nersessian, Nancy (2008). *Creating Scientific Concepts*. Cambridge, MA: MIT Press.

Nersessian, Nancy (2022). *Interdisciplinarity in the Making: Models and Methods in Frontier Science*. Cambridge, MA: MIT Press.

Nguyen, C. Thi (2020). *Games: Agency as Art*. New York: Oxford University Press.

Nickel, Philip J., and J. Adam Carter (2014). "On Testimony and Transmission." *Episteme* 11 (2):145–155.

Nozick, Robert (1981). *Philosophical Explanations*. Cambridge, MA: Harvard University Press.

Oreskes, Naomi (2019). *Why Trust Science?* Princeton, NJ: Princeton University Press.

Peirce, Charles S. (1986). "How to Make Our Ideas Clear." In *Writings of Charles S. Peirce: A Chronological Edition* (vol. 3, 1872–1878, pp. 257–276). Bloomington: Indiana University Press.

Perry, John (1979). "The Essential Indexical." *Noûs* 13 (1):3–21.

Plato (1974). *The Republic*. Indianapolis: Hackett Publishing Company.

Porter, Theodore (1996). *Trust in Numbers*. Princeton, NJ: Princeton University Press.

Potochnik, Angela (2017). *Idealization and the Aims of Science*. Chicago: University of Chicago Press.

Pritchard, Duncan (2008). "Knowing the Answer, Understanding, and Epistemic Value." *Grazer Philosophische Studien* 77 (1):325–339.

Pritchard, Duncan (2010). "Knowledge and Understanding." In Duncan Pritchard, Alan Millar, and Adrian Haddock (Eds.), *The Nature and Value of Knowledge: Three Investigations* (pp. 3–90). Oxford: Oxford University Press.

Putnam, Hilary (1975). "On Properties." In *Mathematics, Matter and Method* (pp. 305–322). Cambridge: Cambridge University Press.

Putnam, Hilary (1978). *Meaning and the Moral Sciences*. London: Routledge & Kegan Paul.

References

Putnam, Hilary (1983). "Models and Reality." In *Realism and Reason* (pp. 1–25). Cambridge: Cambridge University Press.

Quine, W. V. (1980). "Two Dogmas of Empiricism." In *From a Logical Point of View* (pp. 20–46). Cambridge, MA: Harvard University Press.

Quine, W. V. (1990). *Pursuit of Truth*. Cambridge, MA: Harvard University Press.

Rawls, John (1999). "Two Concepts of Rules." In Samuel Freeman (Ed.), *Collected Papers* (pp. 20–46). Cambridge, MA: Harvard University Press.

Rawls, John (2000). *Lectures on the History of Moral Philosophy*. Cambridge, MA: Harvard University Press.

Rawls, John (2008). *Lectures on the History of Political Philosophy*. Cambridge, MA: Harvard University Press.

Raz, Joseph (2012). *The Morality of Freedom*. Oxford: Clarendon Press.

Rechnitzer, Tanja (2022). *Applying Reflective Equilibrium*. Berlin: Springer.

Riggs, Wayne (2003). "Balancing Our Epistemic Goals." *Noûs* 37 (2):343–352.

Riggs, Wayne (2019). "Open-Mindedness." In Heather Battaly (Ed.), *The Routledge Handbook of Virtue Epistemology* (pp. 141–154). New York: Routledge.

Robertson, Emily (2009). "The Epistemic Aims of Education." In Harvey Siegel (Ed.), *Oxford Handbook of Philosophy of Education* (pp. 11–34). Oxford: Oxford University Press.

Rorty, Richard (1979). *Philosophy and the Mirror of Nature*. Princeton, NJ: Princeton University Press.

Russell, Bertrand (1971). "Letter to Frege." In Jean van Heijenoort (Ed.), *From Frege to Gödel* (pp. 124–125). Cambridge, MA: Harvard University Press.

Rysiew, Patrick. (2012). "Epistemic Scorekeeping." In Jessica Brown and Mikkel Gerken (Eds.), *Knowledge Ascription* (pp. 270–294). Oxford: Oxford University Press.

Scanlon, T. M. (1998). *What We Owe to Each Other*. Cambridge, MA: Harvard University Press.

Scheffler, Israel (1960). "Teaching and Telling." In *The Language of Education* (pp. 77–101). Springfield, IL: Charles C. Thomas.

Scheffler, Israel (2008). *Worlds of Truth*. Malden, MA: Wiley-Blackwell.

Sellars, Wilfrid (1968). *Science and Metaphysics*. London: Routledge & Kegan Paul.

Shapin, Steven (1994). *A Social History of Truth*. Chicago: University of Chicago Press.

Sidgwick, Henry (1981). *Methods of Ethics*. Indianapolis: Hackett Publishing Company.

Siegel, Harvey (1988). *Educating Reason: Rationality, Critical Thinking, and Education.* New York: Routledge & Kegan Paul.

Sober, Elliott (2015). *Ockham's Razors.* Cambridge: Cambridge University Press.

Son, Hyungmok, Juliana J. Park, Yu-Kun Lu, Alan O. Jamison, Tijs Karman, and Wolfgang Ketterle (2022). "Control of reactive collisions by quantum interference." *Science* 375(6584): 1006–1010.

Sosa, Ernest (1991). *Knowledge in Perspective.* Cambridge: Cambridge University Press.

Sosa, Ernest (2007). *A Virtue Epistemology: Apt Belief and Reflective Knowledge* (vol. 1). Oxford: Oxford University Press.

Strevens, Michael (2008). *Depth.* Cambridge: Harvard University Press.

Strevens, Michael (2017). "How Idealizations Provide Understanding." In Stephen Grimm, Christoph Baumberger, and Sabine Ammon (Eds.), *Explaining Understanding* (pp. 37–50). New York: Routledge.

Stroud, Barry (1984). *The Significance of Philosophical Skepticism.* Cambridge, MA: Harvard University Press.

Stroud, Barry (2018). "The Epistemological Promise of Externalism." In *Seeing, Knowing, Understanding* (pp. 40–54). Oxford: Oxford University Press.

Suárez, Mauricio (2009). "Scientific Fictions as Rules of Inference." In Mauricio Suárez (Ed.), *Fictions in Science: Philosophical Essays on Modeling and Idealization* (pp. 158–178). New York: Routledge.

Suárez, Mauricio (2024). *Inference and Representation.* Chicago: University of Chicago Press.

Tal, Eran (2011). "How Accurate is The Standard Second?" *Philosophy of Science* 78 (5):1082–1096.

Tal, Eran (2020, fall). "Measurement in Science." In Edward N. Zalta (Ed.), *Stanford Encyclopedia of Philosophy.* https://plato.stanford.edu/archives/fall2020/entries/measurement-science.

Teller, Paul (2018). "Measurement Accuracy Realism." In Isabelle F. Prichard and Bas C. van Fraassen (Eds.), *The Experimental Side of Modeling* (pp. 273–298). Minneapolis: University of Minnesota Press.

Thomson, Judith Jarvis (1990). *The Realm of Rights.* Cambridge, MA: Harvard University Press.

Treanor, Nicholas (2013). "The Measure of Knowledge." *Noûs* 47 (3):577–601.

Tufte, Edmund (2001). *The Visual Display of Quantitative Information.* Cheshire, CT: Graphics Press.

References

Tversky, Amos, and Daniel Kahneman (1974). "Judgments Under Uncertainty: Heuristics and Biases." *Science* 1985(4157):1124–1131.

USA Diving (2019). *Competitive and Technical Rules*. https://teamusa.org/usa-diving/resources/rulebook.

Vaihinger, Hans (2009). *The Philosophy of "As If."* Mansfield Center, CT: Martino Publishing.

Van Fraassen, Bas (1980). *The Scientific Image*. Oxford: Clarendon Press.

Van Fraassen, Bas (1989). *Laws and Symmetry*. Oxford: Clarendon.

Van Fraassen, Bas (2008). *Scientific Representation*. Oxford: Clarendon Press.

Vermeulen, Inga, Georg Brun, and Christoph Baumberger (2009). "Five Ways of (not) Defining Exemplification." In Gerhard Ernst, Jakob Steinbrenner, Oliver Scholz (Eds.), *From Logic to Art* (pp. 293–319). Frankfurt: Ontos.

Vogel, Jonathan (1990). "Are There Counterexamples to the Closure Principle?" In M. D. Roth and G. Ross (Eds.), *Doubting* (pp. 13–27). Dordrecht: Kluwer.

Watson, James, and Francis Crick (1953). "A Structure for Deoxyribonucleic Acid." *Nature* 171 (4356):737–738.

Weatherson, Brian (2008). "Deontology and Descartes' Demon." *Journal of Philosophy* 105 (9):540–569.

Weinberg, Steven (1992). *Dreams of a Final Theory*. New York: Pantheon.

White, Morton (1965). *Foundations of Historical Knowledge*. New York: Harper & Row.

Wilczek, Frank (2016, July 5). "How Feynman Diagrams Almost Saved Space." *Quanta Magazine*.

Williams, Bernard (1985). *Ethics and the Limits of Philosophy*. Cambridge, MA: Harvard University Press.

Williams, Bernard (2002). *Truth and Truthfulness*. Princeton, NJ: Princeton University Press.

Williamson, Timothy (2000). *Knowledge and Its Limits*. Oxford: Oxford University Press.

Wittgenstein, Ludwig (1947). *Tractatus Logico-Philosophicus*. London: Kegan Paul, Trench, Trubner & Company.

Zagzebski, Linda (1996). *Virtues of the Mind*. New York: Cambridge University Press.

Zagzebski, Linda (2012). *Epistemic Authority*. Oxford: Oxford University Press.

Index

Acceptability, 9–12, 14–15, 28, 38, 50, 52, 82–91, 155, 187, 223, 226, 285, 300

Acceptance, 11–12, 35–38, 51, 68, 84, 119, 141, 196–209, 217, 223, 291, 296

  with reservations, 91–95, 100, 120

Accountability, 123, 308

Accuracy, 6, 263, 273–274, 297

Action, 30–32

  pattern of, 216–217

Adaptations, epistemic, 7–8, 17

Adequacy, 6, 62, 84, 95–107, 263, 269

  empirical, 109, 213–214, 245

Adjudication, 50, 53, 81, 96–108

Adler, Jonathan, 40, 153, 211, 221

Aesthetic, 283–303

Agency, epistemic, 3–5, 7–8, 12–17, 37–38, 223–240, 242, 254–256, 303

Albert, David, 294

Alief, 305–306

Ambivalence, 68–69

Approximation, 63, 83, 133, 264, 311

Arbitrariness, 83, 130–131, 187, 224–226, 290–291, 295

Aristotle, 191, 204–205, 294

As if, 279

Assertion, 74, 141–145

Astrology, 104–105, 133, 161–162, 258–259, 308

Audi, Robert, 213–214

Authority, 163–165, 167–173, 308

Autonomy, epistemic, 12, 16, 25, 27–57, 105–6, 121, 138–139, 219–220, 303

Ball, Philip, 297

Barnett, Zach, 195, 199

Barrotta, Pierluigi, 73, 134–136

Batterman, Robert, 270–271

Beauty, 285–286, 289

Beebe, Helen, 197

Belief, 18, 32–35, 37, 40–41, 63, 167–176, 193–194, 214

  aim of, 34–35

  Cohen's sense, 11, 35–36

  disagreement insulated, 195

Bell, Clive, 283–284

Berker, Selim, 12

Brun, Georg, 102–107

Capriciousness, 66

Carelessness, 66, 69

Challenge raising, 65–6, 112, 127, 133–134, 174

Chang, Hasok, 6, 14, 30, 91, 95–107, 253–254, 268

Charity, principle of, 136, 203, 215

Chronicle, 224–227, 292

Claim right, 144–161

Code, Lorraine, 27

Cognizance, 39–42, 45–46

Cohen, L. Jonathan, 11, 35–36

Coherence, 50, 72, 88, 101, 130, 285–286
Commitment, epistemic, 193–197, 207–209
Community, epistemic, 3, 7–8, 12–13, 16–17, 54–57, 69–70, 93, 121–139, 157
  membership, 126–128, 134
Competence, 19–2161–65, 106, 164, 178–189, 202, 231
Concession, 178
Conciliation, 178–179, 181–183, 189, 206
Confidence, 34, 121, 182–183, 206, 260
  entitlement to, 260–261, 298, 303
Conscientiousness, 61, 65–70, 106, 202
Consequentialism, epistemic, 169
Consistency, 37, 50, 72, 96–99, 130
Conspiracy theories, 60, 128–131
Construction, epistemic, 5–8, 13, 241–262
Control, cognitive, 17–25
Conundrums, 113–115
Cooperation, 59, 121–122, 146, 156–160, 259
Correction, 8, 14, 87, 119
Co-tenability, 37, 101, 246
Counterfactuals, 21–22, 238, 256–257
Christensen, David, 179
Critical thinking, 13, 34, 35, 43, 50, 71, 161, 223, 227, 233–234, 239

Data, 106, 224, 267–268
Davidson, Donald, 31–32
Defeater, 141–143, 173, 212, 214, 221, 309
De-idealization, 252–253
Deliberation, 31
Della Rocca, Michael, 294
Dellsén Finnur, 204
Demonstration, 20–21, 23–25, 274–276
Dennett, Daniel, 271
Denotation, 274, 276
De Regt, Henk, 126, 294
Descartes, René, 5, 27–28, 194
Desire, 31–32

Dewey, John, 109, 112, 139, 222
Dieks, Dennis, 214
Difference maker, 228, 269, 276–279, 296, 311
Dirac, Paul, 285
Disagreement, 177–209
  in philosophy, 189–209
Discipline, 218–220, 224–235, 283–284
Dogmatism, 116, 119, 132, 181, 182, 208
Dormandy, Katherine, 170, 173
Doyle, Yannick, 278
Duplicitousness, 74
Dynamics, epistemic, 3, 11, 15, 81–120, 126, 230
  internal, 274–276

Ecology, epistemic, 3–4, 17, 241, and passim
Education, 16, 33, 56, 174, 187, 193, 212–240
Elegance, 111, 284, 285–286, 297–298
Elga, Adam, 171
Ellis, B. D., 251–252
Emotion, 32
Empathy, epistemic, 76–79, 157–159
Empiricism, constructive, 109, 213–214, 245–246, 252, 284–286
Endorsement, reflective, 15, 36–38, 43–52, 68–69, 71, 76, 115–116, 161–162, 223, 230, 238–240, 303
Ends
  epistemic, 36–37, 53, 90–91, 174–175
  final, 31–32
  realm of epistemic, 13, 52, 55, 69–70, 121–139 (*see also* Community, epistemic)
Engineering, conceptual, 102–105
Entitlement, epistemic, 60–61, 141–142, 149, 212
Equilibrium, reflective, 12, 16, 43, 81–95, 115–116, 120, 196, 230–239, 288, 291, 300, 303

Index

Error, 15, 46–48, 70, 74–75, 87–93, 100, 119–120, 125, 127, 132–137, 139, 149, 153, 177–199, 207, 230
   reversible, 71–72
Evidence, 20, 60–63, 65, 72, 84–86, 88, 151–152, 158–159, 169, 175–176, 218–219, 228–229, 233, 236–237, 260
   underdetermination of theory by, 244–245
   weight of, 172, 184
Evidentialism, 39–40, 173, 197, 221, 239–240. *See also* Internalism
Exemplification, 24, 119, 218, 275–278, 281, 297–298, 311
Expectations, 63, 65, 122–123, 126–127, 148, 190
Expertise, 15, 42, 56–57, 62–63, 73, 111, 134–138, 162, 163–176, 178, 285
Explanation, 19–25, 89–90, 110, 225–230, 237, 244, 260, 265, 286, 291–293
Extensions, abundant, 241–244 249, 279
Externalism, 221. *See also* Reliabilism

Fact, stubbornness of, 5–6, 14, 17, 103–104, 255–257, 262
Factivity, 109
Fact-stating sentences, 216–217
Fallibilism, 87, 119, 181, 188
   focused, 188, 198, 208
Falsehood, felicitous, 83, 228, 263–281, 302
Feldman, Richard, 175
Feynman, Richard, 287
Fickleness, 66, 68
Firth, Roderick, 12
Foley, Richard, 180
Form, significant, 283–284, 293
Framing quandaries, 114–116
Frankfurt, Harry, 32
Free and equal, 137

Frege, Gottlob, 207
Fricker, Elizabeth, 221
Fricker, Miranda, 134–135, 199
Friere, Paolo, 219–220
Fruitfulness, 10–12, 102, 112, 115, 119, 188, 203–204, 279
Fumerton, Richard, 63–4, 193, 194
Fundamentalism, 117

Galilei, Galileo, 4, 295
Game, 5, 18, 35, 124–125, 256–259, 262
Gardiner, Georgi, 76
Gatekeeper, epistemic, 287–303
Goldberg, Sanford, 63–65,122–123, 128, 190, 194–195
Goldman, Alvin, 39, 166, 174, 194, 212, 221
Goodman, Nelson, 50, 86, 190
Grasp, 17–25, 280
Grasswick, Heidi, 27
Greco, John, 143, 217–220, 224–225, 234–235
Grice, Paul, 24, 59–60, 64, 78, 145, 157–159, 278, 306
Grimm, Stephen, 8–9, 17, 21–22, 42, 82, 110, 223, 228

Hacking, Ian, 243–244, 255, 267, 310
Hannon, Michael, 83
Harris, Paul, 239
Hawley, Katherine, 63, 307
Heisenberg, Werner, 285
Henderson, David, 180
Heteronomy, 38, 54
Heydrich, Wolfgang, 190
Hills, Alison, 18–25, 110, 223, 280, 305
Hiring practices, 191–192, 205–206
History, 219, 224–227, 292
Holism, 42, 85–91, 172–173, 236
Huemer, Michael, 180
Hughes, R. I. G., 268, 274, 276–280
Humility, epistemic, 181
Hunter, Michael, 126

Idealization, 11, 37, 133, 252–254, 264–265, 280, 302
Imperative
  categorical, 50–55
  epistemic, 53–54
Inconsistency, 96–102
Individualism, epistemic, 27–29
Individuation, 248
Indoctrination, 233
Inertia, epistemic, 88, 106, 117–118, 300
Inference, 14, 20–25, 28, 47, 49, 62–66, 83–86, 92, 100, 116, 222, 262, 294
  ampliative, 14, 20–24, 229–231
  license, 274–276
  nontrivial, 10–11, 42, 142, 169, 222–224, 229–232, 237–238
  style of, 116, 184, 217
Injustice, epistemic, 134–137
Inquiry, end of, 104, 109, 139, 264–266
Instrumentation, 4, 6, 13–15, 85, 95–102, 266–268
Interdependence, epistemic, 16, 27, 29–30, 54–57, 138–138
Internalism, 173, 221. See also Evidentialism
Interpretation, 13, 157–159, 274, 276–277
Intersubjectivity, 15, 51, 121–122, 159, 218, 254–255, 259–261
Invariance, structural, 289–290
Iteration
  epistemic, 7–8, 95–109
  mathematical, 100
Ivanova, Milena, 285, 288, 301

Jäger, Christoph, 76, 167
James, William, 185–186
Justification, 9–10, 43, 82–87, 89–92, 121–122, 138, 160–162, 171, 196–197, 221, 229, 236, 239–240, 258–260

Kant, Immanuel, 13, 38, 50–55, 123, 193, 215
Kelly, Thomas, 178–180
Khalifa, Kareem, 9, 82, 110, 228
Kitcher, Philip, 187–188, 192, 197, 208
KK thesis, 40
Knowledge, 2, 8–9, 39–41, 62–63, 82–83, 183, 191, 212
  testimonial, 142–162, 212, 216
Knowledge-first epistemology, 39–40, 307
Knuuttila, Tarja, 268, 274, 277, 280
Kripke, Saul, 116, 182, 208, 307, 308
Kuhn, Thomas, 109–110, 113
Kvanvig, Jonathan, 41

Labor, division of cognitive, 7, 67, 138, 197–198
Lackey, Jennifer, 212–216, 232, 239, 309
Ladyman, James, 285–286
Leibniz, Gottfried, 49
Leveraging, epistemic, 4, 16, 84, 90, 91–105, 107–109, 118, 224, 235–238
Lewis, David, 45, 191–192, 197, 200–201, 205, 207, 249, 251
Limitations, epistemic, 1–8, 11, 29, 53, 81, 117, 132–137, 265–266, 270–272, 294
Lipton, Peter, 23–23, 110, 252
Literature reviews, 191, 205
Logical space, 184
Longino, Helen, 70, 72, 137
Lopes, Dominic, 289
Luminosity, epistemic, 39–45, 49

Malfatti, Federica, 76, 213–214, 239, 308, 310
Massimi, Michaela, 73, 261
Mathematics, 22, 30, 113, 124, 131, 266
Maxim
  Gricean, 59–61, 64, 66, 67–68, 74–75, 78, 145, 156–159, 278, 306
  Kantian, 51–52

Index

McAllister, James, 287
McGowan, Mary Kate, 243, 251
Measurement, 6, 14, 30, 85–87, 159, 252–253
Methodology, 14, 36, 85–87, 118, 228–229
Mill, John Stuart, 201
Miracle argument, 252
Model, 9, 11, 24, 37, 62, 89, 111, 133, 135, 228, 263–281
  banking, 219–220
Monduschi, Eleanora

Nagel, Thomas, 39
Narrative, historical, 224–227
Negligence, 66–67
Nersessian, Nancy, 24, 73, 111, 136, 273, 223
New questions, 99–102, 112
Nominalism, 200–201, 241–262
Norm, epistemic, 7, 13,6,
Normativity, epistemic, 13, 49–51
Norm-stating sentences, 216–217
Nozick, Robert, 299–301

Objections, raising, 16, 54–55, 69, 121, 191
Objectivity, 6, 14–15, 43, 97, 99, 242, 247, 257–258, 307
Ockam's razor, 246, 293–294
Open-mindedness, 61, 70–73, 118, 182, 188–189, 202, 208
Opinion, second, 170–172, 175
Oppression, 219–220
Optional stop rule, 298–301
Options, epistemically acceptable, 184–189
Oreskes, Naomi, 91
Ought implies can, 147–148, 152, 169

Pattern, 89–90, 95–96, 216–217, 265, 270–272, 280, 296, 311
Peirce, C. S., 139

Perry, John, 47–49
Perspective, 51–54, 76–78, 134–136
Pessimistic meta-induction, 11, 116, 194–195
Philosophy, discipline of, 63–64, 173, 180–209, 214–215, 222
Phobia, 32–33
Plato, 33, 193
Plurealism, 246
Poincaré, Henri, 286
Porter, Theodore, 15, 107
Potochnik, Angela, 265–266, 270–271
Practice, 5, 123–129, 131, 146–147, 151, 164–165, 259, 265
  good in, 125–127, 258–259
  good of, 125–127, 258–259
Preemption, 117, 167–174
Prejudice, 32–33
Pritchard, Duncan, 17, 143
Privilege, epistemic, 243–244, 248–254
Problems, 114–115
Professing, 143, 220
Progress, 95, 108–120, 155, 197
Promising, 144–162, 258–259
Proxy, 92–94, 136
Putnam, Hilary, 139, 252, 261, 296, 299
Puzzles, 112–113

Quandaries, 1, 112–115
Quine, W. V., 87–89, 300

Randomness, 130–131
Range, epistemic, 122, 144, 199–202, 265–266
Rationality, 32–33, 182
Rawls, John, 125, 215, 258
Raz, Joseph, 170–171
Realism, 13, 109, 214, 242–250, 252, 257, 280, 284–286, 296, 310
Reason, 15, 65, 66–68, 130, 148, 160, 169–70, 175–176, 223, 236–237, 290
Rechnitzer, Tanja, 86
Re-engineering, conceptual, 102–107

Reflection, critical, 34–35, 37, 223, 239
Refraction, 247–248
Reliabilism, 9–10, 13, 39–40, 128–129, 160–161,169, 172, 196–197, 212, 239–240, 307. *See also* Externalism
Responsibility, epistemic, 27–28, 33–43, 51, 59–79, 105–107, 137–138, 147, 153, 158, 160–161, 182, 202–209, 223, 232
Revision, 11, 87–89
Revocation, mechanisms for, 132–133
Riggs, Wayne, 70–73, 153, 186
Rorty, Richard, 8–11
Royal Society, 126–129
Russell, Bertrand, 207
Rysiew, Patrick, 83

Scanlon, T. M., 65–66, 223
Scheffler, Israel, 216–220, 246
Self-awareness, 48–49, 64–65, 78, 161–162
Self-confidence, 64
Self-deception, 32–33, 75–76
Sellars, Wilfrid, 104, 139
Shapin, Steven, 126
Sidgwick, Henry, 179
Siegel, Harvey, 223, 233, 239
Simplicity, 284, 285–286, 293–296
Sincerity, 61, 73–76, 106
Skepticism, 13, 28, 153, 214, 245
Sober, Elliott, 293, 296
Sosa, Ernest, 40, 61–62, 194
Speculation, attitudinal, 194–195
Stance
    agential, 3, 11–12, 15, 25, 35, 48–50, 280–281
    doctrinaire, 116–118
    judicious, 116–118
    spectatorial, 3, 9–12, 35, 46–48, 242, 280–281
Standard, epistemic, 5–8, 9, 13, 36, 64–65, 72, 85–87, 151–152
Steadfastness, 178, 180–183, 207–208

Stereotype, 32–33, 103
Strevens, Michael, 9, 89–90, 228
Stroud, Barry, 9, 13
Suárez, Mauricio, 263, 268, 274, 280
Subjectivity, 14–15, 43, 50, 51, 92, 95–97, 106, 121
Suspension, 38, 70–73, 93, 103, 181–183, 200–201, 206
Symmetry, 284, 289–291, 301
Systematicity, 111, 123, 260, 284, 292–293

Tal, Eran, 6, 252–253
Taxonomy, 17,103–107, 230–231, 251, 302
Teaching, 29–30, 54, 211–240
Teller, Paul, 252, 264–265
Tenability, 37, 87, 84, 196–197, 291, 300–301
Testimony, 18, 28, 29–30, 54, 141–161, 163, 168, 176, 211–213, 217, 220–222
Thomson, Judith Jarvis, 144–162
Threshold, 3, 37, 41–42, 63–5, 69, 82–87, 186–187
Trade-offs, 111, 199, 201, 207, 241, 263, 294
Treanor, Nicholas, 109
True enough, 9, 144, 150, 152–153, 154–155, 235, 284
Trustworthiness, 2, 7, 59, 63, 78–79, 126–129, 138–9, 150, 219, 307
Truth, 8, 34–35, 37, 62–63, 74, 109, 128, 154–155, 191–192, 246–248, 261–262, 284–285
Truth-conduciveness, 128, 259–260, 284, 286–288, 295–296
Truthfulness, 60–74, 126–129, 306
Tufte, Edward, 75

Understanding, passim
    advancement of, 11, 128–129, 175–176, 279

Index

331

objectual, 17, 22–25, 41, 42, 109
propositional, 17–25, 41–42
subliminal, 40–45, 305–306
Uptake, 137, 142, 144, 155–156, 216,
220–221, 223

Vaihinger, Hans, 279
Vajont tragedy, 134–136, 308
Van Fraassen, Bas, 86, 199, 214, 244–245,
252, 289
Variable, hidden, 290, 298, 299
Vermeulen, Inga, 275
Virtue, epistemic, 71,74, 78, 244,
259–260, 285–286
Virtue epistemology, 10, 13, 39–40, 307
Vogel, Jonathan, 207
Warrant
augmentation, 100
transmission, 100, 141–142, 145, 149,
161–162, 212, 309

Weatherson, Brian, 36
Weinberg, Steven, 293, 299
White, Morton, 224–227, 292
Wiles, Andrew, 113
Williams, Bernard, 5–8, 74–75
Williamson, Timothy, 39–40
Wilczek, Frank, 287
Wittgenstein, Ludwig, 30

Zagzebski, Linda, 61, 167–176